Death and the Body in the Eighteenth-Century Novel

Death and the Body in the Eighteenth-Century Novel

Jolene Zigarovich

PENN

UNIVERSITY OF PENNSYLVANIA PRESS

PHILADELPHIA

Published by
University of Pennsylvania Press
Philadelphia, Pennsylvania 19104-4112
www.upenn.edu/pennpress

Printed in the United States of America on acid-free paper
10 9 8 7 6 5 4 3 2 1

A catalogue record for this book is available from the Library of Congress.
ISBN 978-1-512-82377-6 (hardcover)
ISBN 978-1-512-82378-3 (Ebook)

In loving memory of Mabeth ("Dolly") Bamberger

CONTENTS

ILLUSTRATIONS

Death and the Body in the Eighteenth-Century Novel

INTRODUCTION

If one were to compile a list of British eighteenth-century novels in which the protagonist has an extended, sentimental deathbed scene it would include few—such as *The Man of Feeling* (1771) and *Tristram Shandy* (1759)—especially when compared to a list of nineteenth-century novels. The heroines in Frances Burney's and Ann Radcliffe's novels all survive their exploits and travails, and Jane Austen's novels, for further example, do not include the death of a major character. In fact, one novel seems to have such a cultural purchase on the "good death" in the period that it drains the impact of subsequent death scenes in fiction, even spawning parodies and satires. Surely, this would be Samuel Richardson's *Clarissa* (1747–1748). Numerous death scenes in eighteenth-century fiction are actually off-scene, or prior to the novel's plotting. As with Henry Fielding's *Tom Jones* (1749), Frances Burney's *Evelina* (1778), and Charlotte Turner Smith's *Emmeline* (1788), the period is littered with novels centered on orphans who have experienced the loss of their parents before the novel's opening pages. (Though some, such as Tom and Evelina, journey to discover their true parentage and that at least one parent is still living.) While this book isn't a study of deathbed scenes, it seeks to excavate the import of death and the body in fiction and, in particular, to demonstrate the power of fleshly artifacts and secular memorials previous to and often quite distinct from a Victorian cult of mourning. Without extensive deathbed and funeral scenes, these are often embedded in fiction and have received relatively minimal critical attention. To provide a material framework and narrative, each of my subsequent chapters commences with scenes from Richardson's *Clarissa* that depict literal and rhetorical relics of death: wax, heart, skin, bone, bowel, hair. With this, I argue that as science sought to expose the mysteries of the body in the Enlightenment, the corpse became an emotionally valuable, cherished relic in fiction. And as anatomists dissected, undertakers preserved, jewelers encased, and artists figured the corpse, so too the novelist portrayed bodily artifacts. My larger aim is to demonstrate the eighteenth century's growing acceptance of contact with human remains in a post-Reformation,

Enlightenment culture, an aesthetic practice largely ignored in criticism and that I claim is, amber-like, immortalized in eighteenth-century fiction.

Why are these morbid forms of materiality entombed in the novel? This book attempts to address this complex question. Foremost, it suggests that the body itself—its parts, or its preserved representation—functions as a secular memento, and it suggests that preserved remains became symbols of individuality and subjectivity. Therefore this project compels us to reassess the eighteenth-century representation of and response to the dead and dead-like body, and its material purpose and use in fiction.

While there are lengthy death studies of the Medieval, Early Modern, and Victorian eras, death in eighteenth-century Britain has not received the same attention. The last three decades have provided noteworthy surveys of English death, such as the work of Ralph Houlbrooke, Ruth Richardson, Esther Schor, Clare Gittings, Peter C. Jupp, Nigel Llewellyn, and Julian Litten, along with historian Paul S. Fritz's work on the eighteenth-century funeral and undertaking trade.[1] But a dedicated volume of the history of death in the eighteenth century is sorely lacking. It is for this reason that historical facts concerning funerary practices and the understanding of this culture's overall relationship with mortality are only beginning to be compiled and assessed. I remedy this critical neglect in part by foregrounding the political work and materiality of dead bodies and uncovering the ways in which fiction embeds cultural attitudes toward death and dying. Scholars such as Elisabeth Bronfen (*Over Her Dead Body: Femininity, Death and the Aesthetic*, 1992), Janet Todd (*Gender, Art and Death*, 1993), and Terry Castle (*The Female Thermometer: Eighteenth-Century Culture and the Invention of the Uncanny*, 1995) have cogently addressed death and representation in eighteenth-century literature and culture. Todd's work on heroic deaths and suicide revealed the cultural fascination of death that she argues is encoded as masculine. Kelly McGuire's *Dying to Be English: Suicide Narratives and National Identity, 1721–1814* (2012) speaks to the public and state dimensions of suicide and the medicalization of death in the period, exposing how the novel analyzes and even contains suicide (committed mostly by male characters). In turn, these studies have typically focused on reading death as a cultural symptom of feminine constructions, a form of self-expulsion from signification, or as a metaphor for textuality itself. Readings of death and representation often rely on Freudian pathologizing, which can certainly contribute to the sources of repression in literature, but my aim is to trouble this psychological component through the historicizing of death and to demonstrate the cultural paradigms underlying this eighteenth-century literary phenomena. Historically, the rituals associated with dying, death, and mourning are gendered (women prepare the corpse and go into mourning, men oversee last rites and burial). While it is beyond the scope of this project to fully interrogate these roles, gender is a necessary element of this study. For instance, subsequent chapters complicate and build on these traditional roles, asserting that preserving the body does become a widespread practice shared by the genders: male anatomists, female wax sculptors, male undertakers, and female hairwork artists all participated. And in no other genre than fiction do we get a full, intimate portrait of corporeal preservation (literal and figurative) and commemoration.

Other notable studies of art-historical cultures of death in eighteenth-century England have informed this study. Nigel Llewelleyn's *The Art of Death: Visual Culture in the English Death Ritual c.1500–c.1800* (1991) records little-known mourning objects, monuments, funereal ephemera, and art, arguing that these objects were part of a rich and varied visual culture of death. Matthew Craske's *The Silent Rhetoric of the Body: A History of Monumental Sculpture and Commemorative Art in England, 1720–1770* (2008) helps situate my reading of funerary art and how it influences fictional depictions. I am indebted to Craske's earlier work on Egyptiana, which helps support my claim in Chapter 2 that preservation practices were growing in popularity. The work of Clare Gittings on the social history of death in England has been indispensable to my overall claim about the rise of individualism and the resulting secular and aesthetic approach to death and dying in the period. Since my argument is centered on the tension between sentimental and commodified mourning practices, I am also indebted to work on death and materiality such as Elizabeth Hallam's and Jenny Hockey's, *Death, Memory and Material Culture* (2001) and Maureen Daly Goggin's and Beth Fowkes Tobin's *Women and the Material Culture of Death* (2016), both of which provide diverse and broad surveys of materiality across periods and cultures. Crystal B. Lake's *Artifacts: How We Think and Write About Found Objects* (2020) examines speculative reconstructions of found matter and how they are imaginatively repurposed in popular culture. Her approach discusses how the decayed, incomplete artifact invites us to create histories.[2] Lake deploys Bruno Latour's assertion that nonhuman objects have agency (now commonly described as "new materialism" theory). Many of the objects I examine do not necessarily have the fragmentation and political power Lake bestows on the artifact, but some do have political investments and, in the least, demonstrate a privileging of matter (a mind-matter Enlightenment dualism that Latour emphasizes but that recent studies have questioned).[3]

Marcia Pointon's work on British hair jewelry and portraiture, as well as Ariane Fennetaux's study of mourning jewelry, assists my larger claims about sentimentality, luxury debates, and the aestheticization of mourning practices. Deborah Lutz's engaging *Relics of Death in Victorian Literature and Culture* (2015) traces mourning artifacts and their significance to Victorians. Her curatorial approach affirms that objects held memories and narratives and uniquely represented the personal and individual. *Death and the Body in the Eighteenth-Century Novel*'s historical remit is distinct and deeply works with the history of the novel and the origins of keepsake practices. While Lutz sees an inevitable connectedness between Victorian relic culture and "the relics of the Catholic cult of saints," I acknowledge this historical connection for the eighteenth-century relic but argue that it has a secular, aesthetic function instead. In her recent study *Narrative Mourning: Death and Its Relics in the Eighteenth-Century British Novel* (2020), Kathleen M. Oliver also supports a secular relic culture with a nuanced, curatorial approach. Her project argues for the relic's role as narrative strategy, a symbol of fictional and cultural losses, stating that "reliquaries contai[n] the consciousness of the (fictional) dead" (16). She deploys Freud's conceptions of mourning and melancholia as she analyzes characters' melancholic and traumatic experiences. Oliver's psychological analysis is compelling, locating the relic's function as a replacement for the dead and its association with memory and loss. While we examine a similar topic, my reading is strikingly different as it argues with numerous examples that the relic represents the slippage between corpse and treasure, sentimentality and materialism, and corporeal fetish and aesthetic accessory.

One of the larger aims of this project is to disrupt traditional readings of eighteenth-century Britain's "rather restrained mourning characteristic" (to use Peter Walmsley's phrase), which is often contrasted with "the more intimate and intense expression of grief that emerged with the Victorians."[4] In fact, my research confirms that archives continually speak to the period's rising funeral and mourning culture, as well as the increasing commodification of death and mourning typically associated with nineteenth-century practices. By incorporating a variety of historical discourses—such as wills, undertaking histories, medical treatises and textbooks, anatomical studies, philosophical treatises, and religious tracts and sermons—*Death and the Body in the Eighteenth-Century Novel* contributes to a fuller understanding of the history of death in the Enlightenment. It is anchored by my reading of fiction as a unique and distinct space to articulate the changing culture: first, from a theological conception to a medicalization of death and the body, and second, from mourning rituals governed by the church to the aestheticization and commodification of practices primarily governed by the individual. While critics have observed that by midcentury death and suicide were firmly established as central features of plot and thematic content of the novel, "which similarly absorbs a cultural interest in melancholia and mourning," I specifically demonstrate the prolific material, corporeal effects of this absorption (McGuire 17).

In the following pages, novels are read as a means to understand the dead and dead-like body in the culture they reflect. As we come to better understand the history of death and mourning practices in the period, we find a means to better understand literature and, more specifically, the novel. I acknowledge that these significant cultural shifts appear across all literary genres but argue that narrative space enables the full and detailed exploration of death, dying, and mourning practices. No other genre would have the capaciousness for Clarissa's death preparation, for instance, or the long-awaited unveiling of *Udolpho*'s wax figure. *Death and the Body in the Eighteenth-Century Novel* deeply examines how anatomy, undertaking, and public displays and circulation of the body infiltrate narrative plots and aesthetics. I claim that unlike poetry and other genres, the novel encases, displays, and circulates bodies or objects symbolizing the dead. As I demonstrate how a growing secular death informs the culture and consequently fiction, this does in part involve the practice of material culture analysis. What role do dead bodies play in the novel? Why is there a desire to preserve their parts? How is this impetus for preservation figured in the novel? How does the dead body function as both a commodity and a sentimental object?

To address these questions, foremost, this study is informed by a literary-historical methodology. In some fashion, I deploy the novel in order to reassess aspects of historical conditions that engendered these shifts in the period. Though the primary ideas of this book in terms of attitudes toward death and the body have roots in ancient practices, the project's focus is the eighteenth century. I am trying to demonstrate that excavating and collating representations of death and its associated material cultures in the novel contributes new and unique insights to the historical narrative. By charting the complexities and paradoxes of mourning objects in the novel, we can illuminate a prolific, diverse, unique mourning culture. Since the midcentury *Clarissa* is a literary touchstone throughout the book, I explore early fiction in order to demonstrate the origins of Richardson's themes and the novel's materiality before examining a wide range of later eighteenth-century narratives. In *Falling into Matter: Problems of Embodiment in English Fiction from Defoe to Shelley*, Elizabeth Napier recognizes "a novelistic lexicon of corporeal signs is rapidly developing after Richardson."[5] And with the cult of sensibility and the rise of Gothic fiction peaking later in the century, the project admittedly reaches forward, making connections to Romantic era practices and fiction.

The body and corpse have an obvious place in the Gothic imaginary, which my study acknowledges. Gothic and sentimental novels were directly influenced by "graveyard poetry," the mid-eighteenth-century movement that meditated on the transience of life, inevitability of death, and consolation afforded by the Christian afterlife. We clearly see the movement's influence across genres and the incorporation of graveyard and macabre motifs. Mark Sandy's *Romanticism, Memory, and Mourning*, which focuses on a variety of poetic forms, speaks to the importance of poetic treatments of loss and mourning meditations.[6] For much of the eighteenth century, the morbid sensibility of the graveyard school of poets showed continuous popularity and influence. Especially in the 1740s and 1750s a new type of writing emerged centered on mourning and sepulchral morbidity. Works such as Thomas Parnell's "A Night-Piece on Death" (1721), Edward Young's *The Complaint: or, Night-Thoughts on Life, Death, & Immortality* (1742), Robert Blair's "The Grave" (1743), which extols a Calvinist Presbyterianism, Thomas Warton's *The Pleasures of Melancholy* (1747), and Thomas Gray's "Elegy Written in a Country Churchyard" (1751), and prose counterparts to the poetry such as Elizabeth Rowe's *Friendship in Death, In Twenty Letters from the Dead to the Living* (1728), James Hervey's *Meditations Among the Tombs* (1746), and later, George Wright's *Pleasing Melancholy: Or a Walk Among the Tombs in a Country Church Yard* (1793), establish the period's discourse about death and mourning. Graveyard poetry depicted a form of mortuary sentimentalism and reveled in the taste for religious verse that facilitated subjective and affective responses to dying, death, and the afterlife.[7] Eric Parisot convincingly argues that midcentury graveyard poetry "occupied a distinct literary niche fashioned by the transformation of the funeral sermon at the pulpit, and by its diminished occurrence in print" (18). Protestant doctrine denied mourners a connection with the dead, thus graveyard poetry became a new creative medium "by which to express the desire for conversation with the absent dead—for both the poet and the reader" (Parisot 18). Here the embracing of death and religiosity converge; we see tensions between spirituality and memorialization as sensibility provides a focus for personal grief. For example, in *Night-Thoughts*, individual contemplation and solitude are rewarded in heaven. Young describes,

> Who think it solitude, to be alone!
> Communion sweet! communion large and high!

> Our reason, guardian angel; and our God!
> . . . .
> How dreadful, *then*, to meet them all alone,
> A stranger, unacknowledg'd, unapprov'd!
> (3:8–10; 13–14)

Here private solitude "grants access to the divine self and heavenly fellowship."[8] Poetry facilitates this solitude, contemplation, and imagination. As with much of graveyard poetry, grief and despair have the potential to connect one with God. Young even includes the discourse of natural science ("particles, as atoms ill-perceived: / As circulating globules in our veins," 6:190–91) in order to demonstrate our ability to reason, while acknowledging these forces emanate from the divine. This starkly contrasts with Gothic and sentimental modes which do not have as central themes divine revelation. For Young and others in the school, death is not aligned with Gothic superstition or horror but with everlasting life and the eternal soul. Blair's aesthetic in "The Grave" is a bit different; reason does not console the fears of bodily death. Death is to be feared: "Oh horrid apparition, tall and ghastly, / That walks at dead of night" (68–69). The corpse and its spirit haunt the graveyard, and the dead body must be restored by the undertaker. For Blair, the living are isolated from the dead ("Tell us ye dead! will none of you, in pity / To those you left behind, disclose the secret?" [432–33]). Yet everlasting life is also the curative to this isolation; in the end, the speaker seeks out scripture and finds solace in the idea of the second coming and the soul's reunion with the body.

In its attempt to remember and speculate about the "unhonour'd" dead, Gray's "Elegy" does embody a Gothic mood (Walpole's "gloomth"): "The ploughman homeward plods his weary way, / And leaves the world to darkness and to me" [3–4]). The forgotten graveyard that houses the dead poor inspires the speaker to elegize. As the speaker navigates how best to remember the nameless dead ("Full many a flower is born to blush unseen" [55]), he simultaneously resurrects them through the imagination. This mode is not theological but more political: the elegy, a genre typically reserved for the notable and glorious, is for Gray refashioned for the rural dead: "Their name, their years, spelt by the unlettered muse, / The place of fame and elegy supply: / And many a holy text around she strews, / That teach the rustic moralist to die" [81–84]). By resurrecting the dead and reading and imagining epitaphs ("*Melancholy marked him for her own*" [120]), the speaker keeps the past alive while reminding those in the present that death is universal and inevitable. He rescues the dead from obscurity and immortalizes them in poetry; words—not images, bodies, or things—do the work of memorialization.

We see this personal and private, yet didactic and public, mode in prose examples as well. Hervey expounds in a footnote, "When we walk over the Dust of our Friends, or kneel upon the Ashes of our Relations, this awakening Circumstance must strike a lively Impression of our own Mortality. And what Consideration can be more effectual, to make us serious and attentive?"[9] For Parisot, the tract implicitly recognizes and "attempts to articulate by a series of affective transactions, the difference between devotion born of fear and self-interest, and an altruistic piety that promotes Christian virtue in others."[10] And I agree that Hervey's tract also marks a step toward the mid- to late-century vogue for sentimental fiction, "a literary paradigm designed to represent and cultivate munificence and sociable virtue."[11] Graveyard literature thus anticipates the two dominant modes: Gothic and sentimental fiction. How it differs is in its devotional context. Hervey's tract as well as much of graveyard poetry strongly reaffirms the Christian commitment to spread the word of God. Parisot reminds us,

> Like the graveyard poems, Hervey's *Meditations* is a transitionary literary text, one that combines Reformation theology with early modern sentimentalism to enable religious devotion. Fear is invoked here as an emotional expression of the desire for self-preservation, but Hervey's proto-Gothic aesthetics also point the way to Horace Walpole's *The Castle of Otranto* (1764), and a Gothic literary tradition that magnifies fear and terror for the pleasure of frisson, rather than piety. Hervey's religious sentimentalism also draws heavily on the grief and sorrow that accompany death as an affective spur to religious action.[12]

Parisot and others recognize that graveyard poetry is underpinned by a deep religiosity, yet it is its sepulchral melancholy that influences primarily secular, aesthetic tropes in Gothic fiction. Thus we can recognize the sepulchral trappings embedded in eighteenth-century fiction as a response to these aesthetics, though noting that they function in quite distinct ways. The graveyard school is theological, aesthetic, and epistemological, "beyond the inertia of the corpse" (Smith 32). Such poetry provides very different contexts in determining how the dead body is imaged and received.[13]

While elegiac poems and pleasing melancholy treatises (Smith's *Elegiac Sonnets*, Warton's *The Pleasures of Melancholy*) depict the tensions between processes of memorialization and the capacity of the imagination to engage with the dead, and clearly and directly influence narrative aesthetic in the period, my focus on the novel emerges from interest in how the relatively young genre develops. Death is written about in new, self-conscious ways, and the emerging novel engenders these multitude functions, allowing us to trace and connect how attitudes toward death, dying, and bodies develop, shift, and modernize. The novel is an ideal medium by which to catalog the extensive materiality of mourning, as well as the reliance on both empirical and subjective experiences. My own emphasis lies in examining how narrative provides a space for detailed, diverse, sensory inspection. Surveying the range of contexts and forms in which the British engaged with circulated, preserved, and cherished bodies helps us identify how and why the novelist incorporated these aesthetic, commodified, and sentimental practices.

Studies such as Andrew Smith's *Gothic Death: A Literary History 1740–1914* (2016) and Yael Shapira's *Inventing the Gothic Corpse: The Thrill of Human Remains in the Eighteenth-Century Novel* (2018) are helpful overviews of death in the period and interpret the corpse as a form of literary entertainment, but as Smith recognizes, "the corpse often refuses to function as a formal Gothic prop" (2). A book that examines the materiality of death in fiction must necessarily address the Gothic, so I do cover the influence of graveyard poetry and the popularity of the Gothic novel. My chapters discuss texts such as Matthew Gregory Lewis's *The Monk* (1796), Clara Reeve's *The Old English Baron* (1777), and Ann Radcliffe's *A Sicilian Romance* (1790), *The Romance of the Forest* (1791), and *The Mysteries of Udolpho* (1794). Yet critical to my argument is that non-Gothic dead bodies often experience scientific, material, sentimental, commodified functions. Though not considered graveyard literature, Samuel Richardson's *Clarissa* is notable as *ars moriendi* and confirms "the fashionable mid-century taste for mournful piety."[14] Anchored by close readings of *Clarissa*, my study attempts to place the eighteenth-century novel within a larger historical framework.

As narrative functioned as a realistic, "to the moment" archive of experience, the body saw increasing relevance as its own source of storytelling. Thus my historical and theoretical approach has been influenced by sensory studies of the Enlightenment such as Barbara Stafford's *Body Criticism: Imaging the Unseen in Enlightenment Art and Medicine* (1991), which explores a remarkable set of body metaphors deriving from both aesthetic and medical practices that were developed during the Enlightenment "for making visible the unseeable aspects of the world."[15] While Stafford shows how these metaphors reflect transformed approaches toward the human body during this period, she also powerfully argues that we need to recognize the occurrence of a profound revolution—a radical shift from a text-based to a visually centered culture. Since Stafford's influential study, others have focused on the other higher senses, troubling a sight-specific reading of Enlightenment culture.[16] I utilize and expand these theories to pivot sensory focus on touch and the corporeal. Specifically, I argue that as death became more personal and secular in the eighteenth century, visual reminders of death rose in popularity, as did tangible, tactile memorial objects. These had previously gone relatively unnoticed, until recent work by Hallam and Hockey, Pointon, Fennetaux, and others. I also employ philosophical treatises on materiality, including the indivisibility of the body and self-touch, that grapple with the apprehension of the solidity of other bodies. This discussion includes theories that rely on the body's absolute presence to itself, making the body available to the touch of the other. My aim is to argue that the body's solidity and preservation become increasingly important in this period as notions of self center on theories of the tactile and material: resistance confirms corporeality. It is this confirmation of bodily existence and presence that transforms into mourning practices that treasure human remains and keepsakes. Surprisingly, hair; tokens; preserved hearts; viscera urns; jewelry; miniatures; wax reliefs; small, tangible, wearable replicas; and relics of the dead signify the materiality of Enlightenment death. *Death and the Body in the Eighteenth-Century Novel* offers ample support for such claims through exploring these and other forms of preservation in fiction.

A further subject that is deeply debated in studies of the Age of Reason that I chart below is the complex role of religion. As the following chapters distinctly argue for a fleshly, secular materiality, they do address religious veneration, post-Reformation practices, and the traces of Catholicism that influence mourning practices. Nancy Armstrong remarks that "The shift from a culture dominated by the church, which also maintains itself by force, to a culture that relies primarily on ideology to keep people in place required no less than a political revolution in many countries."[17] For post-Reformation Protestants, the place of the dead had to be reviewed and renegotiated in light of such a momentous social and theological realignment. Consequently, a topic such as this reveals a complex interweaving of continuities and discontinuities, a coexistence of disparities as violent change impacted religious reformations and the social significance of the dead. In *The Work of the Dead: A Cultural History of Mortal Remains* (2015), Thomas Laqueur states that the aim of the "men of the Enlightenment" was to "take the dead into other uncharted realms: less Reformation than revolution." Anticlericalism was of course one of their motives: "The aim was to deny the church and its priests the power over the dead that they had enjoyed and purportedly abused for so long."[18] And though this aim may not have been aggressively secular—in fact, it did not want to eliminate the idea of a Christian community of the dead—I agree with Laqueur that their aim "was about creating new spaces where new forms of enchantment could flourish" (184). (And *Clarissa* indeed is an example of this more modern sense of the dead body.) Both Michel Foucault and Philippe Ariès claim that the whole notion of death and the dead body having any metaphysical or transcendental importance collapsed in the period. The Enlightenment saw a disenchantment of the dead body as it became just one more locus for the scientific study for life. In *Flesh in the Age of Reason* (2004), Roy Porter remarks that "Enlightened élite found this Christian preoccupation with the flesh—in both punishment and salvation—naïve, vulgar, implausible and gross."[19] Yet in many ways, materialist thinkers indeed had to celebrate the superiority of the flesh, the corporeal over the conscious. And for the anatomist, he punished the flesh maybe not for salvation, but for knowledge. Flesh, bodies, and carnal desires are deemed vile; "flesh was endlessly flayed and punished by the Churches" (472). As they began to replace what Porter terms "old-style Church belief," Enlightenment thinkers gave flesh "a key and honourable function" (472). These notions inform the anxiety about the body and the afterlife and inspire less superstitious ways of coping with death. So, while I acknowledge dying and death were still intimately connected to religiosity, I maintain that this role was being renegotiated in the period. My interest lies in what I argue is a modern attitude toward the dead body, an approach that gets narrativized and archived in a largely secular, empirical medium: fiction.

The traditional perception of the eighteenth-century novel as an instrument of enlightenment has seen recent contestation. For example, Jesse Molesworth argues instead for the novel as an agent of re-enchantment.[20] Though my larger aim is not to attempt to heal the rift between supernatural and secular worldviews, it partly illuminates not a full "re-enchantment" but, in part, a renegotiation or renavigation. As philosophers theorized how we acquire, experience, and process information and knowledge, and empirically explained individual experience, the novelist focused less on the abstract and more on specifics and subjectivity. With this, we can understand that narrative suspense attaches itself to secular, not supernatural, matters. The enlightened, modern reader may be titillated, but not held in suspense, by supernatural events.[21] As I detail in Chapter 1, thus the novel becomes a space for the renegotiation of religiosity and the supernatural; faith is reframed and replaced (with morality, virtue, and other secular concerns).[22]

It is beyond the scope of my project to fully address this debate and its intricacies. As a scholar of the novel, I maintain the seminal views of Ian Watt, Nancy Armstrong, Margaret Anne Doody, Elizabeth Napier, David Richter, Michael McKeon, Helen Thompson, and numerous others that the rise of the novel is one of many developments that mark the cultural move from obsession with religious matters to a growing preoccupation with realism and the present.[23] I am claiming that intense philosophical and medical interest in the body, as well as a growing secular culture, engendered a focus on individualism and the corporeal that necessarily translated into fiction. By asserting this expansion of secularity, *Death and the Body in the Eighteenth-Century Novel* does not cast aside religion in the age. These chapters do claim that religious observance was losing its grasp when it came to concerns over the body and corpse. Noted earlier, this view is grounded by the studies of historians such as Roy Porter and Claire Gittings, who observe the increasing importance of individualism that heightened awareness of the significance and distinction between life and death, and that an expanding consciousness of self led to the desire to mark and mourn loss.[24] As the Western world became increasingly commodified and scientific, constant reference to a transcendent order, the fears of hell, and other religious matters notably decreased.

The main point I wish to emphasize is that these concerns are not strictly supernatural or secular; they are the result of an indeterminate culture poised between competing, contradictory, and mutually informing explanatory systems. Religion and faith persisted as fully secularized subjects were emerging among new revelatory systems of modernity. To borrow Laqueur's wording, this book explores one aspect of a "new form" of negotiation and enchantment: a secular reliquizing of the dead body. Chapter 1 details my reading of an eighteenth-century "beautiful death"—albeit influenced by religiosity and the notion of the Christian good death—hinged on the concept of a rising secular, liberal individualism.[25] As faith was questioned, so too was the understanding of the afterlife. If science explained the corruption of the body, and offered no evidence for the soul, some Christians sought comfort in immortalizing the body. The dying experienced a growing concern about bodily dissolution and the afterlife, thereby preserving the body saw renewed interest. For survivors, possessing remains became both a deeply personal and a public practice, with mainly commemorative (not religious) impetus. *Death and the Body in the Eighteenth-Century Novel* illuminates this complicated dialectic, asserting that by the end of the period, preserving human remains—for scientific, entertaining, sentimental, fashionable purposes—becomes a marker for enlightened cultural formation.

Chapter 1, "Secularism, Materialism, and Death in the Eighteenth Century," examines how these discourses shift control over death and the body from religious institutions to the individual. Working with historians of death and the body, I illustrate that this shift resulted in secular, aesthetic approaches to dying and death. This aesthetic argument is hinged on my reading of the period's growing secularity. In this manner, I disrupt traditional readings of death in eighteenth-century literature that typically focus on Puritan new individualism, a perverse fear of the flesh, and the obliteration of the body (Stannard; McManners). My project uniquely intersects fictional examples with the growing popularity of the undertaking trade and mourning rituals, preservation methods, and relic culture. Janet Todd, Elisabeth Bronfen, Patricia Phillippy, and others have uncovered the ways in which death and femininity are culturally positioned as the two central enigmas of Western discourse.[26] In particular, they represent that which is inexpressible and unmanageable, that which must be controlled by virtue of social law and art. Death and its representation are certainly gendered in Western art and discourse, and my study addresses this complexity, yet my interest is in examining facets of corporeality and the materiality of death at large in order to demonstrate the cultural pervasiveness of the desire to handle, wear, encase, and circulate the preserved body.

To support these claims, Chapter 1 examines influential concepts such as the Christian "good death," the loosening hold of the church, empirical investigations and philosophies, and the cult of sensibility. In my specific reading, the dissected, embalmed, and parceled, the waxed and circulated, the preserved for mourning—all are reduced to pure solidity, productions of a culture centered on empirical evidence. Like empiricism, many novels sought to collect, organize, and process the data of sensory experience. Both inspect with clinical thoroughness the inner workings of human subjectivity. Yet, science and the novel also teach us about the deep history of material fixations. In particular, this chapter interrelates emotions, bodies, and things. To do this, I examine corporeality and philosophies of touch that assist my claim that the desire for tangibility and tactility infiltrates funerary practice. The cult of sensibility's emotional and psychological significance is, conjoined with its reliance on the body and its physical reaction to pain and beauty, one of the many contributing factors to a growing sentimentality in mourning practices that cherished and physically reliquized the lost loved one. This lays the ideological groundwork for subsequent chapters, which demonstrate how sentimentality and the focus on the body engender various forms of secular memorials and preserved remains, relics that embody their own narratives and become symbolic touchstones in fiction.

Chapter 2, "Preserving Clarissa: Eighteenth-Century Embalming Practices," engages English church reorderings and registers, burial and mourning rites, and the growing medical interest in preserving the dead body, an interest that influenced the rise of the undertaking trade. This type of historical evidence supports my unique claim that as a mortuary practice, embalming spread beyond royal tradition (as evidenced in *Clarissa*, which depicts heart and viscera burial) and infiltrated the burial rites of the wealthy and aristocratic as part of an effort to posthumously preserve social distinction. As a result, there was an increasing preoccupation with mortality and the corpse. This revises Julian Litten's argument in *The English Way of Death* (1991), which relegates preservation practices in the era to extreme cases. In addition to the incorporation of urns, heart burial, and other forms of preserved remains in eighteenth-century mortuary practice, the chapter finds that the frequent publication of medical books and treatises compounds the determination that embalming was consistently practiced and discussed, helping to paint a larger picture of death's reception. Embalming practice also demonstrates the increasing preoccupation with mortality and the corpse and the intense anxiety about bodily dissolution and disruption after death. Using *Clarissa* and other novels, I demonstrate a sociohistorical moment in which fiction reflects fact: postmortem preservation was innovated, discussed, practiced, and circulated. It was part of the larger shift in behavior that included using lead coffins, burying in vaults, preserving physical mementos of the departed, and embracing secular funeral arrangements. The chapter closes with the circulation of notorious dead bodies, the curious history of private collections that housed human remains, as well as eccentric cases of individuals requesting the posthumous display of their preserved corpses. Together, I argue that preserving, collecting, displaying, and circulating the dead body reflect a morbid sentimentality directly influenced by empirical thought and experiment, fully fleshed out in fiction.

Chapter 3, "Waxen Encounters: Flesh, Anatomy, and Autonomy," connects ideas of reliquary and embalming from Chapter 2 with another curious form of preservation: the role of wax in the history of *memento mori* and in eighteenth-century medical practices. The closest element to flesh in the period, wax stands in for the dissected body and is central to the anatomists' art, thereby playing an important role in the Enlightenment pursuit of knowledge. I claim not only that wax symbolizes the empirical, tactile shift in the culture and the rise of individualism, but as it straddles religious *memento mori* and secular relic, the waxform loses its former Catholic association. With this, I demonstrate that the role of wax in mourning and funereal practices, though ancient, is refashioned in the eighteenth century. I suggest that in many ways this exposure and preservation of anatomy coincides with the undertaker's modes of preservation and the popularity of memorial tokens made up of human remains or signifying the deceased's body. As a fleshly form of relic, variations of wax and its circulation—as iconography, *memento mori*, mourning seal, waxwork, anatomy model—register the growing cultural shift for personal, tangible engagement with the body and corpse. An ancient material used for effigies, wax is modernized for eighteenth-century mourning purposes: bas-reliefs, seals, and waxworks materialize loss. The translucent, sweaty resemblance to resisting flesh works as a perfect medium not only for circulating houses of waxworks, but for the anatomist, wax modeler, and undertaker. As icons and religious relics are purged from a post-Reformation, Protestant England, the culture seemed to embrace a secular wax medium as a desire for the thingness of the body became popular. Framing the chapter with examples from fiction, I close with an analysis of Ann Radcliffe's *The Mysteries of Udolpho*. Specifically, I argue *Udolpho*'s heroine has a unique somatic response to the wax *memento mori* (that of a rotting corpse), signifying pleasure and autonomy (versus religious affirmation or moral example).

Chapter 4, "Circulating Bodies: Secular Mementos, Jewelry, and Hairwork," and Chapter 5, "Osseous Matter: Bones, Relics, and Mourning Miniatures," gather memorial artifacts that float throughout eighteenth-century fiction, claiming that they receive sentimentalized and even sublime attachments. This circulation of relics points to the shift in mourning practices in the early nineteenth century. As Samantha Matthews, Paul Westover, Deborah Lutz, and others have shown, the Romantic cult of relic making and collecting is well-documented.[27] The afterlives of Percy Shelley's heart, jaw, skull fragments, bone, and ashes, Byron's heart and lungs, and numerous locks of hair from Keats have all been described since their deaths. Yet the role of mourning miniatures, mourning jewelry, and other material memorials in the eighteenth-century novel has received relatively minor attention. *Death and the Body in the Eighteenth-Century Novel* refers to the circulating body parts of the famous and infamous, but with the larger intent to demonstrate the growing acceptability of close contact with the corpse and to include secular reliquaries. In a period that saw new levels of the dissected, displayed, and circulated body, we can see how practices and attitudes in fiction uniquely speak to a culture's engagement with the body and human remains. My extensive archival research includes examples of the circulating dead body (hair, portraits, jewelry, and other mementos), as well as the contents of the curiosity cabinet and anatomy school. These last two chapters claim that these changes resulted in a sensibility toward the dead body and a cherishing of its parts and a desire for mementos that stand in for the body, such as portraits and miniatures, as well as objects owned by the dead. While fear of dissection and concerns about the state of the soul and the afterlife had cultural purchase, a growing individualism—as Richardson details in *Clarissa*—produced an intense concern for how the dead body was treated, revered, and displayed. As this occurs, the state of the mourner is more directly addressed, and the rising mourning industry capitalizes on the desire for mementos. While the Victorians will experience a profound cult of mourning that continues the practice of cherishing human remains and the deceased's treasured objects (though in different forms, as the giving of rings declines, and so on), eighteenth-century culture experiences an intense desire for corporeal contact: wearing and handling mementos of the dead in an attempt to personalize mourning but also to symbolically forestall the dissolution of the body. This refashioning of memorial relics, which I explore in detail in Chapters 4 and 5, speaks to the larger project of fiction: to encapsulate and immortalize the real. It is the novel's emphasis on both materiality and the subjectivity objects can engender that helps form a culture's experience of death and mourning. *Death and the Body in the Eighteenth-Century Novel* seeks to illuminate this material shaping and contribute to the broader conversations about how individual and collective modern formations are narrated.

Throughout the book's chapters I emphasize that the loosening hold of the Church, empirical investigations and philosophies, the cult of sensibility, and rising liberal individualism result in the aestheticization of death and mourning in the period. In my view, this cannot occur without the Enlightenment focus on the body and senses. In *Consumption and Literature: The Making of the Romantic Disease* (2006), Clark Lawlor recognizes that "the more secular concerns of the mechanical and rationalistic philosophy and science of the eighteenth century concentrated more on the body-machine."[28] Studies of the novel such as Napier's *Falling into Matter* have demonstrated the various ways bodily aspects enable and resist the creation of the novel, and how the corporeal is central to the imaginative experience.[29] David McAllister's *Imagining the Dead in British Literature and Culture, 1790–1848* (2018) convincingly argues that Romantic conceptions of the dead advanced and resisted social and political reform, and that the corpse gets reconceptualized in literature as an affective and aesthetic response to the Napoleonic era. Yet these helpful studies do not fully address the impact on narrative of the circulated, preserved, and treasured dead body. As of yet, no single study has collected copious material and literary examples of death and mourning in the period. *Death and the Body in the Eighteenth-Century Novel* not only offers new insights about the effect of a growing secularization and commodification of death on the culture and its productions, but it fills critical gaps in the history of death, using narrative as a distinct literary marker.

Illuminating how eighteenth-century British death culture reformulates a Catholic past during a reformed and revolutionary Enlightenment present is only one project of this book. Pointing to a historical continuity, the afterword argues that many eighteenth-century relic and mourning practices became more widespread and fetishized in the Romantic period, culminating in "the Victorian celebration of death." Yet this embracing of the body experiences notable aesthetic approaches and transformation. Inevitably, technology like photography extinguished the popularity of mourning miniatures, hair jewelry more often becomes replaced by jet, and preserving and cherishing remains takes on a horrid, monstrous connotation in the light of the mutilating effects of industry and war. By the end of the nineteenth century, we see another shift in attitudes about the corpse and death: they are something to be feared, denied, and not embraced.

Through examining the desire for human remains in an enlightened society, *Death and the Body in the Eighteenth-Century Novel* traces only a portion of mourning practices and their translation and significance in fiction. It attempts to avoid sweeping or simplistic generalizations through illuminating the complex layering of approaches to dying, death, and mourning in the period, considers regional and denominational differences, and addresses ideological debates and contradictions in the period. However, at its core, this book is asserting on some level a continuity of cultural practice. Admittedly, it is in no way a holistic or complete study. Much work is needed in the history of eighteenth-century death in England to propel these discussions and debates forward. Ultimately, *Death and the Body in the Eighteenth-Century Novel* asserts a historically and materially inflected interpretation of fiction. An understanding of human conception centered on the corporeal, along with an understanding of the idea of the emerging novel being dependent on a particular conception of the human body, becomes recognizable only in the eighteenth century. This book expands this narrative and cultural history by focusing on the role of the transformed, lifeless human body in generating this discursive practice. In a broader framework, it narrates a fascinating and unique history of the novel that speaks to the cultural formation of modern individualism.

CHAPTER 1

Secularism, Materialism, and Death in the Eighteenth Century

Global engagements and imperial wars, outbreaks of disease, lack of proper sanitation, and a still rudimentary, humoral sense of medicine and human biology contributed to the high mortality rates of eighteenth-century Europeans. While the Victorians would later have a monopoly on the cult of death and mourning, eighteenth-century Britons experienced a burgeoning modern and public integration of death. Women still traditionally cared for the corpse, communities and parishioners cared for and mourned with families, parish graveyards and crypts were built among homes and neighborhoods, deathbeds were surrounded by loved ones, and the corpse was laid out in the home, visited and watched by friends and family. While an uncertainty about the afterlife increased alongside scientific discoveries, burial rites expanded and the preservation of the body received increasing interest. The rise of the undertaker (or "funeral furnisher") occurred in the period, resulting in a growing aestheticization and secularization of death and mourning. The list of mourning and funerary accoutrements for especially the middling and upper classes is extensive: scarves, brooches, rings, miniatures, wax reliefs, armbands, hatbands, gloves, dresses, crape hats, buttons, shoe and knee buckles, toothpick cases, snuffboxes, stationery, funeral invitations and tickets, effigies, death masks, urns, and monuments. In response to the act that disallowed shrouds made from anything but wool, *The Tatler* satirizes excessive mourning dress: "I am further advised, that several of the defunct, contrary to the woolen act, presume to dress themselves in lace, embroidery, silks, muslins, and other ornaments forbidden to persons in their condition."[1] Funeral processions became spaces for the expression of power and prestige and, for some, of fashion. By the end of the period, death was certainly commodified. Matthew

Craske states, "The rituals that immediately followed death slipped into the control of tradesmen."[2] Advertisements for undertakers during the period signal this commodification of death, such as undertaker William Clark's trade card (c. 1739) that lists extensive furnishings both for the home and for dress (fig. 1.1).

The funeral was not solely a rite of religious expression, but one of increasing material and political display. In *The Church of England, 1688–1832*, William Gibson recognizes, "The church's relationship with a variety of trades and crafts was symbiotic."[3] This included being paid by guilds, charities, hospitals, and other

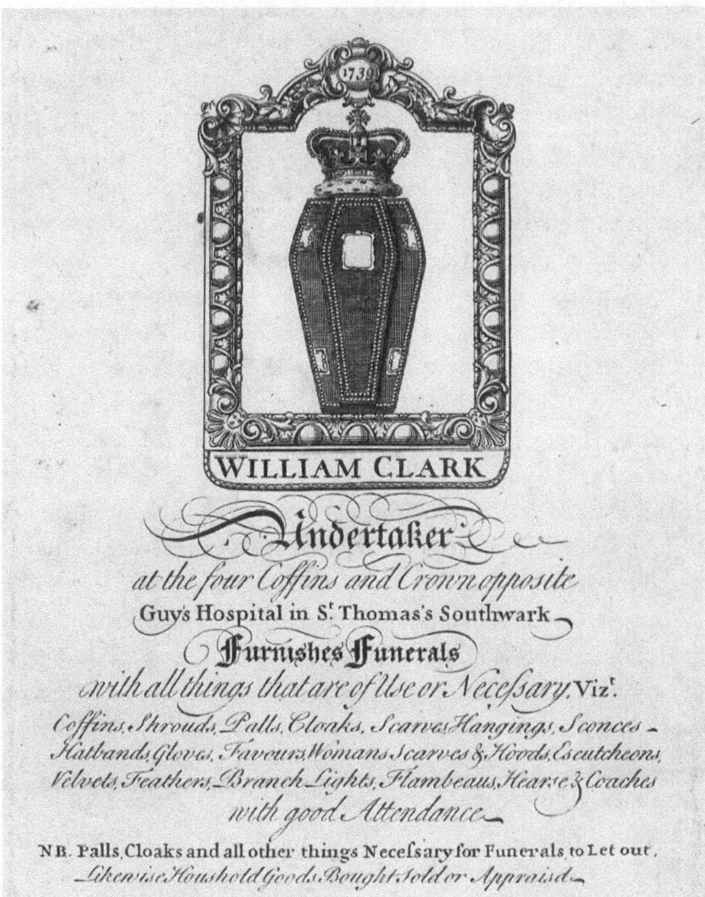

Figure 1.1. Undertaker William Clark's trade card, c. 1739.
© Victoria and Albert Museum, London.

organizations for services, including sermons (with the expectation the sermon would be published and sold). Gibson goes on to explain, "Equally the Church as a consumer fuelled demand for certain goods," creating a market "for ceremonial items (including printed orders of service, funeral directions and monumental masonry), furniture . . . and even clothing (wedding and funeral rings, gloves, scarves and headgear)" (174). For example, the funeral accounts of incumbent of Babbington Hugh Poole (d. 1738) included "payments for a coach and hearse, coffin plate and covering cloth, a crepe suit, fifty pairs of gloves for mourners, and mourning hatbands."[4] This integration of Church and memorial marketing would have the potential to promote and support the religious identity of Anglicanism. On the other hand, as this chapter will chart, commodifying death took ceremony and mourning ritual out of the Church's command.

Control over one's death moved from the Church into the hands of the physician (as in *The Man of Feeling*), but most importantly to the individual (as we will see in *Clarissa*). Advances in medical knowledge led to a broader trust in the profession; for the middle and upper classes especially, the doctor was becoming a more common presence at the deathbed than the clergyman. Indeed, the effects of the Reformation were keenly felt: the role of the clergy in the dying and death process decreased in importance.[5] As the sickroom became the realm of the doctor, the undertaker sought control over the corpse. Craske explains, "The burial books of Westminster Abbey, in which the names of London's undertakers are recorded as they paid the fees, suggest that few persons of 'quality' were, by the 1720s, interred without their services."[6] Having died in debt and estranged from her husband, but still requesting "an earnest wish to be buried in a manner agreeable to her rank," Elizabeth Hamilton Stanley, countess of Derby (d. 1797), was buried once the Hamilton family "discharged her debts, amounting to near 5000l."[7] Her aristocratic funeral is described in detail in *The Gentleman's Magazine*: "Three mourning-coaches and six followed the hearse; in the first was carried the coronet and cushion, all very richly decorated with escutcheons, and other trophies; the next was the Countess's carriage; then twelve others, belonging to different Nobility," who are then subsequently listed (349). Lady Derby's exorbitant funeral illustrates the need for rank to be distinguished by the fashionable funeral, yet by the end of the century the commercialization of mourning creates a blurring of social class. Craske recognizes this shift in the period: "Undertakers were associated with the creation of a fashionable modern world in which all sense of traditional class structure was disappearing" (46). The trade evolved to supplant numerous tradesmen with specialized skills, including surgeons who typically were in charge of embalming corpses. As undertaking became an increasingly necessary trade for a successful funeral, and absorbed the skills of other tradesmen, funerals became more fashionable, as pall, feathers, mourning carriages, and other furnishings were under their provision. This in part homogenized expectations for the funeral of the merchant as well as those for the aristocratic classes. "Undertaking, therefore, was one of the many retail trades that promoted common custom across the wealthy and leisured classes," Craske recognizes (41–42). This *theatrum mundi* was due in part to the converging commercial and sentimental approaches to death.

While etiquette and conduct books will explicitly guide appropriate mourning responsibilities for the Victorians, those in the eighteenth century were still navigating and debating these issues. In his *Tableau de Paris* (1781–1788), Louis-Sébastien Mercier describes in detail appropriate dress and time lengths of eighteenth-century French court mourning. Translated and published in English, the book does not see an equivalent in the period for Britain. Numerous British magazines did publish opinions about mourning practices and fashion, such as *Walker's Hibernian Magazine, Or, Compendium of Entertaining Knowledge*; the November 1786 issue published "Essay on Mourning," which provides global practices and details German mourning dress on the death of a royal (the King of Prussia had recently died) and of particular relatives and describes appropriate dress for half mourning (dress of white and black) along with other protocols. Since the Reformation there had been numerous religious treatises on moderate mourning, such as the anonymous *Preparacyon to Deeth* (1545), but works that detailed mourning protocols and fashion (for those other than royalty and the aristocracy) were few. Instead, directives were embedded in religious treatises and "how-to-die-well" publications. In *The Pathway to Prayer and Pietie Contayning an Exposition of the Lord's Prayer* (1616), the Jacobean preacher Robert Hill lists six reasons to maintain the practice of mourning and the use of mourning wear:

1. By it we keep a memory of our friend.
2. We are drawn to some humiliations.
3. We are put in mind of our own mortality.
4. It argues his love that he bestows it upon us.
5. By this means many poor are clothed.
6. It is but a Legacy of the dead to the living.

He further states that "If the heart mourn as well as the habit, I do not think it utterly unlawful."[8]

Despite the lack of British guidebooks ritualizing eighteenth-century mourning practices, there were consistent modes of treating the dead body and how it was to be treasured and mourned. For example, women were becoming more interested in fashionable mourning habits early in the seventeenth century, as noted by Anne Clifford. She remarks that after Queen Anne's funeral in 1619, "I went to my Sister Beauchamp to shew her my mourning attire."[9] Stages of mourning dress were already in place as diarist Samuel Pepys remarks in the entry for December 8, 1667, which describes that Anne Hyde, the Duchess of York, was in second mourning for her mother: "And so to White Hall, where I saw the Duchesse of York, in a fine dress of second mourning for her mother, being black, edged with ermine, go to make her first visit to the Queene since the Duke of York was sick."[10] Royal protocols were quite elaborate and specific, often set by the College of Arms as well as the grieving family. Even for the death of a monarch, it was only at court that the strictest mourning was observed (mapping French court practices). For example, when Queen Caroline died in 1737, the King ordered public mourning "as for a wife," that is, six months' deep mourning and six months' second mourning, and "all the world cry out upon the hardship of it with respect to trade and private families."[11]

While there was no sole source for British mourning dress and other proprieties, art and literature from the period describe in detail mourning customs and the typical three stages of mourning wear. In Daniel Defoe's *Moll Flanders* (1722), Moll testifies when she goes before a justice that "I was going the Day to buy some Cloaths to put myself into second Mourning."[12] In fact, as a form of *ars moriendi*, novels help us better understand the prevalence of mourning customs and practices. In Frances Sheridan's *The Memoirs of Miss Sidney Bidulph* (1761), types of mourning dress are mentioned for both men and women: when her mother dies, Sidney is in "deep mourning"; the widower Mr. Arnold "la[y's] aside" (129) his mourning attire when he marries Sidney; and Lady Sarah, a cousin and sister-in-law of Sidney's, dons "slight" (307) mourning attire following Mr. Arnold's death.[13] A notable example is also found in Richardson's *Pamela; or, Virtue Rewarded* (1740), which inspired numerous illustrations and paintings, including twelve by Joseph Highmore, painted in 1745. One depicts Pamela, a servant for an upper-class household, in typical mourning for her mistress. In the novel, she writes that Young Squire B "has given mourning and a year's wages to all my lady's servants . . . and he ordered the housekeeper to give me mourning with the rest."[14] In the painting *Mr B. Finds Pamela Writing* (c. 1743), Pamela is shown seated "in my late lady's dressing room," wearing a plain but stylish black robe, tied across the front with black ribbons. Depicted writing a letter, the subject of the novel, and not working, she straddles subservience and leisure. Pamela's cap, a common mourning requirement, is rather fashionable, flat and pinned on top of the head and trimmed with a black ribbon (fig. 1.2).[15] Lou Taylor states that "Due to the rising social aspirations of the lower middle classes, public mourning was now displayed by an even wider social range than in the previous century. Mourning dress was creeping one step further down the social ladder—to the disgust of the higher classes."[16] With Pamela's fashionable mourning, Highmore's painting symbolizes this slippage and signals Pamela's eventual triumph and reward for virtue: her marriage to Mr. B. Since the upper classes found mourning portraits fashionable, their dress needed to symbolize their social and political status. While Pamela's dress signaled her marginal status and ultimate move to the upper class, other paintings affirm social position. A fine example of this is Thomas Gainsborough's portrait of *Maria Walpole, Countess of Waldegrave, Later Duchess of Gloucester* (c. 1763) in black mourning dress (fig. 1.3).

Figure 1.2. Joseph Highmore, from *Four Scenes from Samuel Richardson's "Pamela", I: Mr B. Finds Pamela Writing*, 1743–1744. Oil on canvas. 651 × 759 mm. Photo © Tate London.

Figure 1.3. Thomas Gainsborough, *Maria Walpole, Countess of Waldegrave, Later Duchess of Gloucester*, c. 1763. The History Collection/Alamy Stock Photo.

The quality of her dress (silk and lace), the lack of cap, and her expensive jewelry signify her social standing. An object of beauty, the countess visually defines the aesthetic expectations for an aristocratic woman in mourning.

Charlotte Smith's *Emmeline, The Orphan of the Castle* (1788) details a typical mourning set for a fashionable woman of status: the vain Mrs. Ashwood "finished the important business of accommodating a pile of black feathers, jet and crape, upon her head, 'the mockery of woe' which she did not even affect to feel" (161–62). (With "mockery of woe," Smith quotes from Pope's "Elegy to the Memory of an Unfortunate Lady.") The narrator continues the critical description with, "Mrs. Ashwood, full of her increased fortune . . . busied in studying to make her mourning as becoming as possible" (138). As Lorraine Fletcher mentions, "Formal mourning was elaborate: the crape would hang down in streamers around the wearer's face and neck."[17] Later Mrs. Ashwood mentions putting on "the trimming on the white bombazeen" (a twill material).[18] The narrator's derisive tone clearly signals Mrs. Ashwood's mourning excess and vanity. Apparently, true feelings of mourning and melancholy in the novel are reserved for the virtuous and sensible.

Craske explains that this notion of the fashionable mourner did not go without being criticized: "The rituals of the public death ceremony became so routine, and were so indiscriminately applied, that their ridicule also became commonplace" (43). In his early *Familiar Letters* (*Letters written to and for particular friends, on the most important occasions* . . . [1741]), Richardson closes with three sample condolence letters that seek to moderate mourning. As Walmsley describes them, "These letters warn against an excessive sorrow, which would seem to question God's dispensation, and encourage the use of 'Reason and Piety . . . to rebuke the Overflowings of your Grief.'"[19] This is quite remarkable advice from the man who several years later will pen the most indulgent mourning novel of the period. Despite warnings against excess, "Funerals seemed to have become a province of 'fashion-mongers' who dealt in hastily assembled props" (Craske 46). Satirical articles on modern funerary customs "vilified undertakers for attracting the vain and socially ambitious into ruinous expense."[20] Craske cites a critic's essay in the *Connoisseur* of October 1754 that states, "Such refinements as the giving of mourning rings and commissioning of elaborate stationery seemed to moral commentators to have become meaningless in their promiscuity." The critic goes on to complain that "We cannot suppose that black edged paper was ever intended to be defiled by vulgar hands, but contrived (like gilt paper) for the use of the polite world only. But alas! We must always be aping our betters."[21] Entrepreneurs, this commentator observed, competed to introduce funerary novelties to sell in undertakers' shops. Undertakers, of course, played a role in this modern sense of a fashionable, capitalist mourning industry. As the eighteenth-century English public experienced more capital and goods, the emotional economy of mourning took a new direction.

Allan Ingram and Leigh Wetherall Dickson observe in *Disease and Death in Eighteenth-Century Literature and Culture: Fashioning the Unfashionable* that literature is a product of, and produces a cultural discourse around, "disease and death that intersects neatly with fashion and fashionability."[22] And in no other genre than fiction does this discourse receive a full and detailed treatment and critique. Eliza Haywood's *The History of Miss Betsy Thoughtless* (1751) expresses the debate over sentimentalizing death and mourning. When Betsy's first husband maltreats her and then dies, she performs the rites of widowhood, but admits she does not miss her cruel spouse. When her friend Lady Loveit wonders at Betsy's observing all the proper forms of mourning, considering how glad she must be to be rid of Mr. Munden, she remarks, "'Mistake me not, dear Lady Loveit . . . I do not pretend to lament the death of Mr. Munden as it deprives me of his society, or as that of a person with whom I could ever have enjoyed any great share of felicity, even though his life had made good the professions of his last moments;—but I lament him as one who was my husband, whom duty forbids me to hate while living, and whom decency requires me to mourn for when dead.'"[23] Following Mr. Munden's tragic suicide, Lady Loveit expresses the duties of a widow to Betsy, while warning her against being overly "afflicted": "Decency obliges you to wear black, forbids you to appear abroad for a whole month, and at any public place of diversion for a much longer time; but it does not restrain you from being easy in yourself, and chearful with your friends" (552). When Betsy responds that "there is a shock in death, which one cannot presently get over," Lady Loveit replies, "If we thought too deeply on it, we should feel all the agonies of that dreadful hour before our time, and become a burden to ourselves and to the world" (552). Admitting that "she rather indulged affliction, because she thought it her duty to do so," Betsy agrees that "It was not therefore very difficult to reason her out of a melancholly, which she had in a manner forced upon herself, and was far from being natural to her" (552).

A similar critique of custom versus feeling is made in Sarah Fielding's *The Adventures of David Simple* (1744), where on one of his stays in a lodging, David observes the reaction of three sisters to the death of their father, the inn's owner: "He had three Daughters, every one of whom, attended him with the utmost Duty and Care during his Illness, and at the approach of his last Moments, shewed such Agonies of Grief and tender Sorrow, as gave our Hero great Pleasure."[24] The three daughters "never ceased crying and lamenting, till their Father was buried, in all which time Mr. *Simple* did all he could to comfort them; but, as soon as the Funeral was over, they dried up their Tears, and seem'd quite recover'd" (36). In fact, to David's horror, they argue over their father's belongings. On observing this quick change from grief to avarice, David remarks, "As to their Grief at his Death, there is to most People a Terror and Melancholy in

Death itself, which strikes them with Horror at the sight of it: And it being usual for Families to cry and mourn for their Relations, till they are buried, there is such a Prevalency of in Custom, that is not uncommon to see a while House in Tears, for the Death of those very People they have hated and abused while living, tho' their Grief ceases with their Funerals." David mentions that the "*Ceremony of crying*" is a performance as if "they had loved one another ever so well" (38). This slippage between true feeling and performance, deep grief and custom, is evident throughout the period, including late-century novels such as Frances Burney's *Evelina* (1778), where one of Madame Duval's most unfeeling gestures is giving up her weeds after only three months.[25] This unwillingness to follow proper mourning since it hampers female fashionable dress is evident in numerous fictional examples. In Sheridan's *Miss Sidney Bidulph* (1761), when Sidney's husband dies in a tragic hunting accident, Sidney's pompous cousin and sister-in-law Lady Sarah visits Sidney "in all the formal parade of a state visit" (305). Sidney later remarks in a letter to her friend Cecilia, "How ill does the vanity of pomp suit with a house of mourning!" (305). Lady Sarah—who as a family member should traditionally don mourning attire—is obviously more interested in donning fashionable attire. Sidney is critical of this decision, observing that her sister-in-law refused to wear proper mourning attire: "I found my sister was going to throw off her mourning intirely; that which she had on being so slight, that it was scarcely to be distinguished for such" (307). In Mary Robinson's sentimental novel *The Widow* (1794), the heroine Julia St. Laurence is given advice by her maid, Bernard: "'Besides, you should leave off mourning,' Bernard explains; 'why, no body minds losing a husband now-a-days, and widows never wear weeds above three months.'"[26] Thus novels reflect the debates surrounding proper mourning rituals, depicting excessive mourning practices and criticism of fashionable mourners.

When novelist Frances Burney's beloved sister Susanna died on her arrival to England after her journey from Ireland on January 6, 1800, Burney vowed to keep the date as a day of mourning for the rest of her life.[27] This practice of perpetually keeping a day of mourning would have seemed excessive for some, even earlier in the period. For example, in Lady Mary Wortley Montagu's account of the Austrian court in 1716, she described the constant mourning worn by Princess Christine Louise of Oettingen-Oettingen, the mother of Empress Elisabeth Christine, and noted that "nothing can be more dismal than the mourning here, even for a brother," indicating the existence of different cultural standards.[28] In the same year, Louis XV (or his regent) cut the required lengths of mourning times in half, so that full mourning only lasted six months, "and the rest in proportion."[29] Yet by midcentury, and certainly by century's end, sentimentality had fully entered British burial and mourning practices, especially for the middling and upper classes. Excessive emotion, often aligned with women, was a sign of unreasonable sensibility, so as a result excessive and extensive mourning practices were often criticized, as Inchbald's Lady Loveit recognizes. (For instance, wives were generally not allowed at their husbands' gravesides in order to avoid ungoverned emotions.) Burney's perpetual mourning signals what will become a full sentimental, Romantic mode of mourning at the end of the eighteenth century.

Several recent studies have examined the period's luxury debates and interrogated the economic and moral effects of the production and consumption of luxury goods. Consumer consumption and global, imperial expansion certainly affected discourses on sensibility, taste, and aesthetics. For this study, the imbrication between material culture/sensibility and consumerism/mourning rites is an integral space for exploration. Mourning economies are directly tied to Britain's growing global economies and the desire for exotic new materials and novelties. A fascinating material example of this intersection with mourning practices is found in the print culture following the collapse of the South Sea Company, the British joint-stock company created in 1713 to reduce national debt by investing in the supply of African slaves to South American and the "South Seas." When the stock price reached exorbitant highs following bribes and insider trading, and then crashed in 1720, the economic bubble ruined thousands of investors (Sir Isaac Newton supposedly lost £20,000) and diminished the national economy. Numerous poems and satires were written, and illustrations invoked the crash as a form of national death and mourning. Catherine Ingrassia describes one of James Milner's satires, *Three Letters, Relating to the South-sea Company and the Bank* (1720), which depicts the South Sea Company and the Bank of England as two women and describes the "Lady of the Bank in deep mourning."[30] But the title of this satirical funeral ticket (c. 1721) for the directors of the South Sea Company of bubblers, "Death is the End of all Men," speaks to the intersection of financial speculation and mourning etiquette (fig. 1.4). The ticket invites the holder to "accompany the whole Body of S.S. Directors from ye Bubbling house in the Broad way" in a funeral procession to "ye three Legged Tree" where a funeral sermon will be preached by "P.P.P. who has been a noted sufferer by being two bisie with Capital Stock."

Capitalism and mourning practices converge in often problematic ways. While this book focuses on the body as a form of mourning commodity, it demonstrates that these accessories that stand in for or are constructed from the

Figure 1.4. George Bickham the Elder (c. 1684–1758), *Death is the End of all Men*, c. 1721. Satirical funeral ticket for the directors of the South Sea Company. Etching and engraving. 179 × 223 mm. © The Trustees of the British Museum.

body are often tied to the luxury economy: mourning portraits on ivory, gold mourning rings, diamond mourning buckles and buttons, silk mourning weeds, and expensive black-edged stationery. In *Luxury and Pleasure in Eighteenth-Century Britain*, Maxine Berg recognizes that sentimental funeral presents reflected this new economy that binds families and communities in ways that exchanges of money could not. She writes, "Small items of silverware were vividly described and bequeathed with care among near and distant kin; mourning rings were allocated, and family watches identified and passed down. All of these possessions had functional and symbolic identities connected with social status, period and place, taste and manners."[31] It is no coincidence that the rise of the undertaker coincides with the rise of eighteenth-century consumerism and global trade, as *things* became representative of kinship and community. Mourning clearly became more commodified, its sensibilities spilling over into public performance and social practice. In subsequent chapters, I will discuss how undertakers and entrepreneurs were integral to the practice of preserving, copying, and reproducing the dead body for sentimental mourners.

A large aim of *Death and the Body in the Eighteenth-Century Novel* is to explore the complex intersection of the material and the sentimental, the commodity and the keepsake. In order to provide context for and to support my claims about death and the novel, this chapter will further examine conceptions of the good and beautiful death. This aesthetic argument is hinged on my reading of the period's growing secularity, as I trace the social acceptance of individual control over dying and death. Next, I will discuss empiricism, corporeality, and philosophies of touch, which assist my claim that the desire for tangibility and tactility infiltrates funerary practice. Last, the chapter examines ways in which the body, the cult of sensibility, and the novel converge, underscoring the novel's cataloging of emotional and somatic response. Together, these ideologies form a material and empirical view of the novel; like the curiosity cabinet, the novel houses public rituals and private subjectivities, uniquely offering a portrait of bodies and things, as well as their interplay.

The Christian Good Death and *Clarissa*

The eighteenth century proves a unique period in terms of cultural attitudes toward death and the body.[32] Sermons, handbooks, and funeral and burial practices all demonstrate the increasing preoccupation with mortality and the corpse, and intense anxiety about bodily dissolution and disruption after death. As I will later discuss, a growing secular control over death was seen as the Church's grasp on the body and soul loosened. This tension between traditional Christian doctrine and a new liberal individualism is distinctly reflected in the material and print culture of the Enlightenment. Although there are several early literary examples that include death—rather than marriage—as a climax, for instance, Thomas Southerne's tragedy *The Fatal Marriage; or The Innocent Adultery* (1694) and William Congreve's tragedy *The Mourning Bride* (1697), from midcentury on there appears to be a pronounced persecution of lost virtue.[33] We see the exploitation of the misery of the abandoned sexual victim and, likewise, death for the whore (as vividly illustrated in William Hogarth's *A Harlot's Pro-*

gress [1732]).[34] Paula R. Backscheider recognizes that "From the time of the earliest English novels, writers sometimes used the death scenes of heroines to close their texts, to determine interpretation."[35] Backscheider cites novelist Penelope Aubin's philosophy on this gendered form of narrative closure: "Let [the vicious woman's] Station be ever so great and high in the World, nay, let her Crimes be ever so well conceal'd from human Eyes; yet . . . her Death . . . will be accompanied with Terrors, and a bitter Repentance shall attend her to the Grave: Whilst the virtuous shall look Dangers in the face unmov'd, . . . dying with comfort, be freed from the Miserys of this Life, and go to taste eternal Repose" (59).

In fact, as Aubin observes, we can witness Christian values and beliefs being reassessed in literature, most blatantly in Richardson's novels. While Richardson's Protestant motives are evident and well-documented, I utilize *Clarissa* throughout *Death and the Body in the Eighteenth-Century Novel* as a literary marker for changing attitudes toward dying, death, the corpse, and funerary practice. I have argued elsewhere that Clarissa's preparations for and ultimate death signals a shift from the Christian good death to a secularized and aesthetic model, a claim I wish to further develop throughout this book.[36] In *Consumption and Literature: The Making of the Romantic Disease*, Clark Lawlor agrees: "Clarissa's death moves the discourse of the good and easy death on to a newly aestheticized dimension."[37] These shifts can be traced to the changing religious, philosophical, social, and artistic milieus of the eighteenth century.

Nearly the last third of Richardson's novel is devoted to the preparation and eventual death of Clarissa Harlowe. As discussed above, for the eighteenth-century reader this would not be overly macabre. Ian Watt remarks, "Consequently the scope and importance of funeral arrangements had increased until, by Richardson's day, they had attained an unprecedented elaboration."[38] In the journey toward her demise following her brutal rape by Lovelace, Clarissa surrounds herself with and takes comfort from the accoutrements of death. Mirroring the changing English attitudes toward death, Richardson has Clarissa take part in the joyful acceptance of death and union with her "Heavenly Bridegroom": "I have much pleasure in thinking of death, than of such a husband," she pronounces (6:374, letter 91). Exchanging her white bridal garments, now forever tainted, for the "all-quieting garb" of the shroud, she writes, "As for me, never Bride was so ready as I am. My wedding garments are bought . . . the *happiest* suit, that ever bridal maiden wore for they are such as carry with them a security against all those anxieties, pains, and perturbations" (7:373, letter 96).[39] As a good eighteenth-century heroine should, Clarissa plans a marriage with death to redeem her loss of virtue (though this virtue is erotically enshrouded).

In addition to Clarissa's bridal/burial dress, an elaborately designed coffin is ordered by Clarissa and placed in her bedchamber. She explains her reasoning to Lovelace's friend Belford: "I shall not die the sooner for such preparation—Should not everybody make their will, that has anything to bequeath? And who that makes a will, should be afraid of a coffin?" (7:309, letter 81). Belford is shocked at the intricacies Clarissa has designed her coffin with, questioning, "Is a coffin a proper subject to display fancy upon?" (7:310, letter 81). Belford writes to Lovelace of the delivery of the coffin: "It is placed near the window like a harpsichord, though covered over to the ground: and when she is so ill that she cannot well go to her closet, she writes and reads upon it, as others would upon a desk or table" (7:331, letter 88). The coffin, what Clarissa calls her "palace" and "house," is intricately inscribed with biblical quotes and a painting of a serpent: "The principal device, neatly etched on a plate of white metal, is a crowned serpent, with its tail in its mouth, forming a ring, the emblem of eternity" (7:311, letter 82). By staging her death slowly and beautifully, Clarissa paints a portrait of death that opposes her nightmare of being violently cut and buried among disintegrating corpses. "It is probably with this threatening image of dissolution in mind," states Garrett Stewart, "that she imagines her appearance among her ancestors in the coffin she has meticulously designed for this occasion."[40]

Clearly, the capaciousness of the novel form affords Clarissa ample opportunity to profess and elaborate her approach toward her own death. The scene in the death room paints a portrait of the beautiful death that we will see revisioned in the nineteenth-century novel. Roseann Runte recognizes the significance of Clarissa's courting of death: "Eighteenth-century romanesque heroines, less subject to death by duels and other rapid means of extinction than their male counterparts, generally had more time during the course of their demise to expound upon it."[41] Clarissa chooses to have these objects of death (her burial garments and her coffin), always within her sight, "reading and writing upon this self-designed 'house'" (360). Although she threatens to often, Clarissa's puritanical beliefs will not allow her to commit suicide, so she chooses the next best thing: she perpetuates her own death.[42] Her death preparations aren't all altruistic: they are obsessive and consciously symbolic. Surrounded by mourning friends, Clarissa enacts the deathbed scene beautifully. She tells them, "It is not so hard to die, as I believed it to be!" (8:5, letter 1). With a sweet smile "beaming over her countenance," she blesses and forgives her friends and family, including Lovelace, her rapist. She tells Belford, "Do you, sir, tell your friend that I forgive him! And I pray to God to forgive him . . . Let him know how happily I die—And that such as my own, I wish to be his last hour" (8:5, letter 1). Belford describes Clarissa's joy in death as "her happy preparation, and still happier departure" (8:7, letter 1). As a good Christian, Clarissa claims, "Most happy has been to me my punishment here!—happy indeed!" (8:5, letter 1). At her death, Clarissa's beautiful countenance is recorded in Belford's own "death-bed reflection": "Such a charming serenity over-spreading her sweet face at the instant as seemed to manifest her eternal happiness already begun" (8:7, letter 1). By giving her life willingly, triumphing over sexual violence, and painting herself the picture of Christian forgiveness, Clarissa dies an exemplary good death. This bodily decay, a slow, joyful dying, will later be evident in the self-destructive "illnesses" of Clarissa's nineteenth-century sisters.

The opening of this chapter described the pomp and commodification of the eighteenth-century funeral. Simultaneously, a sentimental, intimate approach to death and mourning emerged, creating slippage between the public and private that this chapter will attempt to illuminate. In *The Female Thermometer*, Terry Castle notes a change in eighteenth-century death and mourning: "Where death was once a spectacle of considerable magnitude, it now became primarily a private event, witnessed by only one's closest relations."[43] By this time, funerals were carried out more and more discreetly. The return of her body to her father's house and the subsequent funeral are all overseen by Clarissa from the grave. Her posthumous letters detail how she would like her corpse displayed. Belford, Clarissa's executor, writes to Lovelace describing the exquisite beauty of the corpse, that Clarissa's hair is set in "crystal," she wears her bridal shroud of white, and "flowers and aromatic herbs" fill the hearse to mask any odor (8:40, letter 15). As I will discuss in Chapter 2, although Clarissa was never embalmed, in death she is aligned with the marker of sainthood: an uncorrupted corpse. On the body's arrival at her father's house, Colonel Morden (Clarissa's cousin and subsequent "funeral journalist") writes a detailed description of the "fair corpse" for Belford, who does not attend the funeral: "The corpse was very little altered, notwithstanding the journey. The sweet smile remained" (8:74, letter 21). Clarissa's family is amazed at the corpse's condition: "She was their very niece," "The same benignity of countenance! The same sweet composure! The same natural dignity—She was questionless happy! That sweet smile betokened her being so" (8:75, letter 21). The viewers of her corpse claim that "they never saw death so lovely before" (8:14, letter 6), that death had not altered her features. Her beauty in death reminds the viewers that Clarissa has transcended human sinfulness; her saintly preservation represents the "good death" to which most Christians aspired. The uncorrupted body becomes a physical symbol of a moral and virtuous life. Friends and family kiss and embrace Clarissa's corpse, and are "attracted by the awful receptacle" (8:83, letter 23)—the decorated coffin. Though on one hand her funeral is private and intimately planned by Clarissa, it is also made public. "A great crowd" of townspeople, Morden describes, gather to gaze at the "coffin-lid, and the devices upon it" (8:88, letter 24). It is a spectacular scene in which Clarissa envisioned the pleasure others would receive from the display of her beautiful corpse. Indeed, her funeral is a scene of writing; it is, as Anna exclaims, "all of . . . Clarissa's story" (8:79, letter 22). Not only the messages painted on Clarissa's coffin but the messages on her face and the condition of her beautifully preserved corpse are read by viewers as signs of happiness and as an assurance of their own existence.

Clarissa, in the lengthy approach to her death, has the time to write letters and her will, which include personal reflections, posthumous messages, and directions for her burial and execution of property and belongings. In this sense, Richardson's novel reflects the growing secularization of the will in the eighteenth century. Historians generally agree that people still strove for a good death in the eighteenth century, often exemplified by Clarissa's intense and detailed preparation (though her good death is aesthetic and commodified, and managed not by the Church but by herself). Eighteenth-century culture's burgeoning individualism, alongside Lockean liberalism, resulted in a segregation of political and domestic domains. Death slowly found itself governed by the individual in a manner quite different from Jacobean or Restoration manifestations. As we will see, the rise of sentiment in the period shifted the rhetoric of one's last moments from fearful and didactic to pleasing and affective. As Walmsley puts it, "Dying well became, more and more, a loving act: to expire with joy, or at least patience and fortitude, served as both an inspiring example and a comfort to the bereaved" (5–6).

The most important sign of the changing sensibilities in the period is the emergence of what historian Philippe Ariès terms a European "cult of the beautiful dead" or what Terry Castle refers to as a "romantic cult of the dead," a growing subjective fascination with idealized images of the deceased.[44] Beautification was used to hide the physical signs of mortality and decay and to overcome any sense of separation for loss of individuation. By 1750, an attitude toward death that Ariès calls "the death of the Other" is securely in place, distinguished by a more secularized philosophy of death. "A new sensibility made the death of loved ones more cruel for survivors," identifies Ariès in *The Hour of Our Death*, "and led to an almost fanatical cult of remembrance." This sentiment "took advantage of the taste for mummies," for embalming and coffining, and for the "custom of preserving a body in the home."[45] As a result, the increased secularization of death produced a new fascination with the body and corpses, spawning scientific research into the mysteries and medicalization of death. Whereas Western man was learning what it meant to be a living body, what it meant to master the body, for some the fear of death diminished. In *The Birth of the Clinic*, Michel Foucault argues that an epistemic shift occurred at the end of the eighteenth century, that a rediscovery was made of the late Renaissance conviction that knowledge is possible on the basis of death. "Death as the absolute point of view over life and opening on its truth," he argues, "is also that against which life, in daily practice comes up against."[46] Yet even as the scientific and philosophic emphasis on rationality tried to combat death by exploration and explanation, a new uncertainty about the status of the body and a return to superstition were equally prevalent. In *Over Her Dead Body*, Elisabeth Bronfen notes that "the eighteenth century marks an increase in necrophilia, in an anxiety about premature burial and a fear of the living or reanimated corpse, the vampire."[47] Thus the status of the body and corpse proves a problematic and troubled cultural space, associated with deeply rooted religious and supernatural convictions.

While the influence of Puritanism simplified funeral rites and resisted sensibility, by the middle of the eighteenth century a complete reversal had taken place. English Puritan funerals were characteristically unceremonious and not costly. Out of fear of emotionally responding to death and incorporating rites resembling papism, Puritans limited their mourning ceremony: funeral sermons were not typically read at the time of burial, flowers didn't adorn the tomb, attendance was often limited to a few family and friends of the same sex as the deceased, and the dead body was not venerated.[48] Yet the rise of individualism and the cult of the beautiful death resisted this Puritan model. According to David Stannard in *The Puritan Way of Death*, "the emphasis was now placed heavily on the individual who had been so fortunate as to die, with little said about the community's loss."[49]

This focus on the individual's religiosity is an important paradigm. Peter Marshall observes, "Before the Reformation, to elucidate the geography of the afterlife was the legitimate business of theologians, devotional writers and homilists. To attempt to do the same in post-Reformation England, however, was characteristically to engage not with a physical, but with a metaphysical question, and increasingly perhaps, a metaphorical one."[50] There was a fundamental reappraisal of the known physical world and the metaphysical conceptions of the afterlife, territories no longer relegated to theologians. In his *Philosophical Letters* (1734), Voltaire, one of the most famous Enlightenment anticlericals, pronounces, "This is a country of sects. An Englishman, being a free man, goes to

Heaven by whatever path he chooses."[51] As Eric Parisot describes, "An ever growing number of divergent religious perspectives coupled with an array of theological issues concerned with the proper way to salvation meant that the pious no longer viewed one straight and narrow path, but a variety of roads that equally claimed salvation as their destination."[52] This liberty of choice inevitably requires a certain degree of self-authority. In his study of religion and graveyard poetry, Parisot cites anticlerical John Trenchard, who summarily captures the overriding contemporary concern over competing authorities on salvation. In *The Natural History of Superstition* (1709), Trenchard asserts, "Since the Divine Providence has for the most part hid the Causes of Things which chiefly concern us from our View, we must either entirely abandon the enquiry, or substitute such in their room, as our own Imaginations or Prejudices suggest to us, or take the Words of others whom we think Wiser than our selves."[53] Here he underscores the battle between organized religion and moral autonomy (and inherently refers to the deists). In a broader framework, anticlerics were highlighting that the competition for religious authority was essentially underwritten by the increasing tendency (albeit "vain" and "enthusiastic") to claim the self as a legitimate source of religious knowledge.

In part to combat these evident changes, in the late seventeenth and early eighteenth centuries, sensibility infiltrated funeral practices, sermons, and consolation guidebooks. Publication of "funeral literature," popular "death-bed reflections," and best-selling theological works dealing with death soared by midcentury. Along with the change of mourning rites from the public to the private sphere, and the increased production of tombstones and memorials, the preaching of preparation sermons testifies to the erosion of older funeral practices as the individuality of the deceased was more often stressed. The publicized sermon became a literary vehicle of resistance to the growing secularity of death and dying. Pastors reacted by publishing sermons full of warnings against not preparing for death properly. Puritan homiletics in the seventeenth and eighteenth centuries used the fear of hell and the decay of the body as the main weapon against these cultural shifts. Confessing sins, praising God, and asking forgiveness were part of the necessary requirements to avoid hell. As this puritanical mode gradually diminished, though, preachers turned to popular "how to die well" books for inspiration; these texts place significance on the Christian "good death" and avoid instilling graphic fear in the reader. The unimaginable joys of heaven are instead emphasized. This mode is recognized in Catholic practices as well. In *Death and the Enlightenment: Changing Attitudes to Death Among Christians and Unbelievers in Eighteenth-Century France*, John McManners remarks, "In the best writings on death, hope and consolation outweigh the burden of judgement and fear. In them we find the two traditions of evangelism in tension: the tradition of the pulpit and the tradition of the confessional . . . or in spiritual letters, countering, or sublimating the fears which the thunders of the sermons excited."[54] Several sermons incorporate the consoling language of early works, such as Charles Drelincourt's *Christian's Defence* (1651) and John Kettlewell's *Death Made Comfortable* (1695). According to Stannard, "It was becoming the norm, the accepted norm, for the godly to die 'in Raptures of holy Joy'" as the formal doctrines of Puritanism were withering (150). The change in rhetoric is noticeable in these more consolatory sermons. No longer are the fires of hell and the corrosion of the body referenced; death is portrayed as a marriage and directions are given so that we can enter the eternal world, or as minister Daniel Turner phrased it, "with safety."[55]

A part of this "safety" involved receiving deathbed rites. In this same period, we see the continued popularity of Jeremy Taylor's *The rule and exercises of holy dying* (1651), which encouraged the "sacramental" tradition with the Church of England, emphasizing the desirability of making a confession of sins to the clergyman, receiving absolution and the last Communion (viaticum). As Houlbrooke notes, "The English Catholic community maintained as far as possible the ancient last rites of their faith, including extreme unction as well as absolution and the viaticum, and continued to support masses and prayers for the dead."[56] Julie Rugg observes that a growing medicalization of death emerged: "As a greater understanding of medicine and the use of opiates removed terror of the arbitrary incidence of death and the deathbed agony, Christian teaching on the afterlife could take a gentler tone."[57] For many, the God of the Enlightenment was a rational being, "distant, benevolent, tolerant and unmysterious" (Rugg 204). This benevolent image did not correlate well with the notion of eternal damnation and the pits of hellfire, and this aspect of the afterlife is softened (or even avoided) in consolatory literature published in the second half of the century. As noted in numerous examples of deathbed scenes in the Methodist *Arminian Magazine* (a periodical read by as many as 100,000 people by the end of the eighteenth century), "To die a good and holy death, free from anguish and uncertainty, was the aspiration placed before the Methodist faithful." David Hempton states in *The Church in the Long Eighteenth Century*, "The experience of death is recounted in language replete with metaphors, this time of slumber, surrender, flight, departure and rest in the arms of Jesus. Also present is resistance to Satan and his evil legions. In death evil is resisted and defeated, eternity is welcomed and embraced."[58] Foremost, the stories described "earthly renunciation and heavenly anticipation in which lives well spent in disciplined holiness were given the ultimate reward of triumphant glory" (158). Dissenters such as Elizabeth Singer Rowe, author of the didactic romance *Friendship in Death: In Twenty Letters from the Dead to the Living* (1728), imagined a detailed portrait of heaven: "The chrystal Gates open'd a spacious Entrance, and the blest Immortals received me to the Mansions of Life and Bliss."[59] Heaven is a "fragrant," "enchanting Land of Love" with "delectable Vales and flow'ry Lawns." Backscheider recognizes that more than any other early writer, "Rowe developed the death scenes of women into a powerful sign system that went beyond the exemplary women characters' beautiful deaths and the fallen dying terrified and desperate" (59).

In Catholic Europe, the doctrine of Purgatory offered a potential path of comfort: purgatory was a concept that allowed for the working through of one's sins with the hope to eventually enter heaven. The removal of purgatory and prayers interceding on behalf of the dead post-Reformation reduced the power of the priest as giver of the last rites "and increased that of the family and friends" (Lawlor 37). Together with the adoption of the doctrine of predestination, "all served to undermine the lengthy round of services and masses for the dead associated with burial in the pre-Reformation Church. Such elaborate rituals were rejected as superstitious and superfluous; the doctrine of predestination taught that the consequences of death and the achievement of salvation could not be altered though human actions."[60] Thereby, simple burial services markedly contrasted with Catholicism's external observances. Walmsley describes, "The Catholic Church stressed the efficacy of intercessions by the living—especially prayers and masses—in shortening a soul's time in purgatory" (6). Protestant theology does not offer this balm; the soul directly journeys to either heaven or hell. To confront the "softening" of the horrors of hell seen in many of the period's sermons and consolatory literature, we see sermons such as John Wesley's "Of Hell" from October 10, 1782, written in the Church of England tradition with Puritanical detail of the horrors of "the lake of fire burning with brimstone" and "exquisite torment of body, and horror of soul."[61] Natural philosophy, skepticism, and sensory, empirical evidence will trouble this confidence in the Christian afterlife.

The funeral also became a pulpit for the preacher: he used the death of another to remind survivors that they must follow particular steps in order to ensure a good death and afterlife. Thus the sermons became guidebooks for dying properly and implied that without these provisional steps, we are all at risk of living in eternal fire. Treatises such as Reverend Zachary Taylor's *The Decency and Moderation of Christian Mourning* (1702) sought to control the outward display of grief. Yet, as we have seen, by midcentury Puritan influence on mourning (in terms of stark ritual) was losing its hold, and the scope and importance of funeral arrangements had increased, even seeing an unprecedented elaboration, foreshadowing the Victorian cult of mourning. Funerals and burials dramatically changed in both form and content to meet the needs of a more secular society. The funeral service became, in Ralph Houlbrooke's terms, primarily "a vehicle of instruction for the living."[62] The funeral sermon, still strongly tied to religious belief, dwelled on the life of the deceased, gave assurance of his or her life in the hereafter, and consoled survivors. According to Patricia Whiting, "The funeral sermon became the most important feature of the service," and these sermons were not always confined to funerals but were of course printed and sold to the public.[63] This move from resisting to embracing sensibility is reflected in these sermons. Stannard remarks, "The shift here from fearful anticipation to eager longing for death . . . runs through virtually all the available materials on death and dying during this period, from poetry to sermons, from journals to sepulchral art."[64] For example, Craske notes that by the 1730s there was a demand for monuments without "ghastly" elements, and that the hourglass, skull, and crossed bones are removed by midcentury (48). (Chapters 4 and 5 discuss this change in iconography in more detail.)

In the eighteenth century, death ceased to be the ultimate enemy; we see a transition from it being understood in purely religious terms to death being thought of in a more personal light. Foucault claims that at this time, "death becomes the most secret aspect of existence, the most 'private.'"[65] The ultimate form of exercising power over the body turned from a political source (which threatened the living body) to an individual one. Optimistic philosophies and more sensible sermons and treatises eased the horrors of death and moderated invocations of the "King of Terrors." Lawlor effectively describes this notion of a good, gentler death, "An idea of a more gentle and benevolent God gave people a feeling of greater security about a happy afterlife: death could now become a gentle friend and its process a soft transition like going to sleep rather than a struggle with Satan" (Lawlor 40). In *English Society in the Eighteenth Century*, Roy Porter notes that "Plenty of Christians still saw the grave as the gateway to salvation, but others faced dying in new ways" (298). As many people sought private means of mourning, some traditional public forms of mourning inevitably declined. As Walmsley recognizes, "The funeral sermon and the elegy, which flourished in the seventeenth century, were increasingly seen as moribund and wholly conventional" (10). Rituals became more organized and systematized. Sensibility, a focus on grief and loss experienced by the bereaved, and death used as a moral tale for the living were noticeable. This distinct change and Christianity's struggle to focus on the preparation for death and hopeful salvation in the afterlife is interestingly revealed in sermons and devotional tracts.

Following the devotional, *ars moriendi* tradition that attempted to soften the intensity of death, the best of these sermons emphasize hope and consolation versus the burden of judgment and fear. Sermons provide evidence that the church attempted to hold on to its intercession with the person and death in the face of publications that placed emphasis on the repentance of the individual, such as J. Stephens's *A gold chain of four links, to draw poor souls to their desired habitation . . . With some necessary directions to die well* (c. 1701), John Kettlewell's *Death made comfortable: or the way to die well. Consisting of directions for an holy and an happy death* (1695), William Sherlock's *A Practical Discourse Concerning Death* (1705), John Hayward's *The Horrors and Terrors of the Hour of Death and day of judgement, that seize upon all impenitent and unbelieving sinners. With holy directions to die well* (1707), and most notably, the translation of Charles Drelincourt's *The Christian's Defence Against the Fears of Death, With seasonale directions how to prepare our selves to die well* (1651; English ed. 1675). Books on how to die well peaked in the seventeenth century, but the eighteenth century saw a steady stream of translations of Drelincourt's *The Christian's Defence*, which had for many years been enormously popular in England and on the Continent. As its title suggests, the book intended to allay its readers' fears of death by making death a state of existence to be longed for rather than dreaded. Filled with beatific personifications of death as the blessed "Heavenly Bridegroom," it contains erotic descriptions of the dying as willing virgins stripped of their earthly garments.[66] As this sensual passage demonstrates, no longer could reason conquer the dilemma of death, and accordingly, as Clare Gittings has observed, the sermon had to soothe instead of frighten. The melancholic direction of contemplating and embracing death is found in religious works such as editions of the nonsectarian Taylor's *Holy Dying* (a tract with medieval Catholic roots that Lovelace reads in *Clarissa*), which advises that we "make Death as present to us, our own Death, dwelling and dress'd in all its Pomp of Fancy, and proper Circumstances,"[67] and Thomas Stapleton's *Prayers and pious considerations for every day on the week, to obtain a happy death* (1753), as well consolatory literature such as James Hervey's *Meditations among the tombs* (1746) and George Wright's *Pleasing melancholy or a walk among the tombs in a country church yard* (1793), a later example of melancholic consolation which contributes to the development of Romanticism.

The persistent publication of "pleasing death" sermons and handbooks that delineate "how to die well" clearly demonstrates one facet of shifting attitudes toward death, the body, and the afterlife. Rugg explains that "By the end of the eighteenth century, a combination of Enlightenment theory, medical advances and neoclassical aesthetics had begun to rob death of both its terror and its Christian symbolism. The move toward a more secular experience had begun, but was by no means a smooth transition" (209). The hellfire and damnation preaching that characterized evangelical revivalism restoked the notion of eternal torment (well into the Victorian period), though they were powerfully combated by strong sentiments attached to pleasing death and heavenly reunion. Aesthetic, beautiful, and pleasing, dying and death were increasingly guided by influences outside of the Church. A notable example of this affective shift is exemplified by Adam Smith in *The Theory of Moral Sentiments* (1759), which describes an empathetic death:

> We sympathize even with the dead, and overlooking what is of real importance in their situation, that awful futurity which awaits them, we are chiefly affected by those circumstances which strike our senses, but can have no influence upon their happiness. It is miserable, we think to be deprived of the light of the sun; to be shut out from life and conversation; to be laid in the cold grave, a prey to corruption and the reptiles of the earth; to be no more thought of in this world, but to be obliterated, in a little time, from the affections, and almost from the memory, of their dearest friends and relations. Surely, we imagine, we can never feel too much for those who have suffered so dreadful a calamity.[68]

For Smith, thoughts of death awaken compassion versus fear, deeply influencing his theory of sensibility.

The polemical debates, evaluations, denunciations, and reassessments of the Enlightenment seem to have existed since its intellectual infancy. It can be generally agreed, though, that in the tide of the Glorious, Agricultural, Scientific, and emerging Industrial revolutions we see the birth of a modern culture. In my view, the period's shift of power from priest to physician to the individual in managing death is one of many signs of the period's secular turn. Scores of historians and religious scholars have noted that in the Enlightenment and the "Age of Reason," "religion and churches played an increasingly marginal role in political, cultural, and social life—all this leads some scholars to see the period as the start of secularization."[69] For some, there has even been a divide between moderate and radical (hostility to religion) interpretations of Enlightenment secularization; the period contained so many diverse ideas, scholars tend to speak of multiple Enlightenments.[70] The overarching model of the period's more moderate secularity was formed by seminal studies such as Alan D. Gilbert's sociologically informed *Religion and Industrial Society: Church, Chapel and Social Change, 1740–1914*, which argued that urbanization and growing industry challenged the power of the Church of England. For Gilbert and others, the existence of dissent and Methodism perhaps are testimonies to the failure of the Church of England in the pastoral sphere. According to Phyllis Mack, the Enlightenment has traditionally been seen as "the cradle of modern secular society when the religious culture of the pre-modern era was fatally undermined by the new mechanical philosophy and the corrosive criticism of the *philosophe*."[71] In *The Secular Contract: The Politics of Enlightenment*, Alex Schulman argues that the period experienced not only the separation of the political-temporal sphere from spiritual matters, "but the application to politics of a new, evolutionary form of knowledge birthed from the scientific revolution, a form of knowledge that of necessity diverges from both divine-scriptural authority and the hierarchical-dogmatic authority of a priestly caste."[72] For Schulman, "Solidifying progress presupposes religious disestablishment . . . the timeless perfections of divinity and scripture have to be *intellectually* disestablished," and this is necessarily accomplished via science and secularization (2). Historian Margaret C. Jacob's recent *The Secular Enlightenment* underscores these common readings of Enlightenment ideas and practices being a point of departure for the secular world: "It did not necessarily deny the meaning or emotional hold of religion, but it gradually shifted attention away from religious questions toward secular ones."[73] Jacob's study is global, examining how the Americas, Asia, and Africa enter European consciousness through art and trade, and how urban spaces perpetuated discourses and a growing consciousness of "a vastly altered spatial universe" (20). In her secular reading, Enlightenment science leaves the issue of the relationship between God and the universe, between spirit and matter, between space and bodies, "profoundly unsettled."

Despite these recent diverse and global studies, we must acknowledge, along with Gregory and numerous others, that the Enlightenment's secular narrative is currently being rewritten. Phyllis Mack notes this change: "The current upsurge of religious belief and practice in many parts of the world has led contemporary scholars to re-think the secularization paradigm and to approach the subject of religion and the supernatural in a radically different way; not as a relic of the pre-modern era but as a constitutive element of modernity" (86). For example, Knud Haakonssen's collection *Enlightenment and Religion* and Brian Young's *Religion and Enlightenment in Eighteenth-Century England* were part of numerous studies in the 1990s that examined the alliance between religion and Enlightenment thought. Part of this ongoing reassessment necessarily involves returning to the religious views and belief systems of Enlightenment thinkers. It has been well documented that early proponents of the "Scientific Revolution" were convinced that its teachings provide more potent proofs of the existence of the Christian God and might even help restore unity to Christendom. As Stephen Toulmin explains, "The seventeenth-century founders of modern science and philosophy had theological commitments which shaped their whole enterprise. Repeatedly, Descartes and Newton express concern about the religious orthodoxy of their ideas."[74] Teleological beliefs were argued in such well-known texts as William Derham's *Physico-Theology* (1716) and William Paley's *Natural Theology or Evidences of the Existence and Attributes of the Deity* (1802) and were mentioned by many empiricists such as David Hume in his *A Treatise of Human Nature* (1739) and later in *Dialogues Concerning Natural Religion* (1779). In the second edition of his *Principia*, Isaac Newton remarks in a well-known passage that "This most elegant system of the sun, planets, and comets could not have arisen without the design and dominion of an intelligent and powerful being."[75]

The view that God is revealed in his works and that an intelligent creator deliberately designed the natural world would be held by many in the period. As Stewart J. Brown and Timothy Tackett explain in their introduction to *Enlightenment, Reawakening, and Revolution, 1660–1815*, "The revelation of science . . . confirmed the revelations of Scripture, while at the same time ensuring a more rational grounding for the Christian religion, based on the universal laws of nature." Thus ethical precepts of Scripture and the laws of Nature emanated from the same source: a benevolent God. Brown and Tackett assert that "Most of those who embraced the ideals of the Enlightenment remained within the Christian tradition, viewing human reason and natural laws as aspects of God's created order, and seeking to bring a more moderate and ethical spirit to Christianity."[76] These views of a majority Christian moderatism are well supported. In *Miracles in Enlightenment England*, Jane Shaw argues that a larger range of commentators were able to balance "religious enthusiasm" with "reason." In her view, elements of the supernatural evident in an Enlightenment worldview clearly challenge older models of an Enlightenment hostile to religious sensibilities.[77] Gregory illuminates this balance in his reading of John Wesley as an example of the "seeming ability to hold together faith and reason . . . seen as part of a wider pattern of the age."[78] Ulrich L. Lehner agrees, remarking that "Today, historians recognize that only a small fraction of Enlighteners was anti-religious; the overwhelming majority was interested in finding a balanced relationship between reason and faith."[79] Likewise, in *The Religious Enlightenment: Protestants, Jews, and Catholics from London to Vienna*, David Sorkin seeks to reframe the traditional view of the Enlightenment as the starting point of modernity and secular rationalism and argues that most of the Enlightenment's participants were able to unite reason and faith: "The Enlightenment was not only compatible with religious belief but conducive to it."[80] With this, Sorkin rewrites the master scholarly narrative of a "unitary secular Enlightenment project" (3). Historians now talk of a "Catholic Enlightenment," study the practice of Judaism and Islam in Europe, and assess religious toleration in the period. Religion and British colonial relations, the connection between religion and national identity, and revisionist examinations of the Church of England are also fertile areas of study.[81] Many of these studies suggest that Christianity (including Enlightened Catholic discourse) was both compatible with reason and the agent of human progress. Thereby, for certain camps, religious fervor and influence in the period is seeing renewed attention and reassessment.

Materialism, Corporeality, and Touch

The ancient association of sight as a metaphor for knowledge and rationality contributed to the Enlightenment's association with vision. As discussed in the introduction, there have been numerous critical studies of Enlightenment experience that privilege sight. Inevitably, the hierarchy of the senses witnessed persistent discussion throughout the period.[82] The eighteenth century—an age of empiricism—saw understanding the body as central to the science of man. In medicine, literature, and the arts, the theme of corporeality focused debates about correct human responses, expressing emotion, representing beauty, and cultivating relationships. While I do not disagree with the culture's growing reliance on visual culture as philosophies of experience and the significance of sight to human understanding emerge, I wish to pivot this discussion and argue that the body's solidity and preservation become increasingly important as notions of self center on theories of the tactile and material.

In his essays on the "Pleasures of the Imagination" (*Spectator*, nos. 411–21, 1712), Joseph Addison curiously describes sight in terms of touch:

> Our sight is the most perfect and most delightful of all our senses. It fills the mind with the largest variety of ideas, converses with its objects at the greatest distance, and continues the longest in action without being tired or satiated with its proper enjoyments. The sense of feeling can indeed give us a notion of extension, shape, and all other ideas that enter at the eye, except colours; but at the same time it is very much straitened and confined in its operations, to the number, bulk, and distance of its particular objects. Our sight seems designed to supply all these defects, and may be considered as a more delicate and diffusive kind of touch, that spreads itself over an infinite multitude of bodies, comprehends the largest figures, and brings into our reach some of the most remote parts of the universe.[83]

Here Addison is typical in describing sight as "the most perfect and most delightful of all senses," the sense that supplies all the "defects" of the other senses. Yet his analysis of sight as the highest sense is an aesthetic description that relies on touch. Though he analyzes the powers of feeling in terms of its confines and limits, he is incapable of analyzing sight, the "most perfect" of the senses, without invoking the tactile, using such terms as "a more delicate and diffusive kind of touch," "spreads," "bodies," and "into our reach." This is an explicit acknowledgment that the senses do not work in isolation.

In *Strictures on the Modern System of Female Education* (1799), Hannah More (Anglican and Methodist) explains the primacy of touch in an analogy about vanity: "Vanity, if I may use the analogy, is, with respect to the other vices, what feeling is in regard to the other senses; it is not confined in its operation to the eye, or the ear, or any single organ, but diffused through the whole being, alive in every part, awakened and communicated by the slightest touch."[84] Like Addison, she uses "diffuse" to describe the effect of touch, but in her privileging of the sense notes the limits of sight: it is contained to one organ while sight is "alive," "awakened," and "communicated." Edmund Burke's *A Philosophical Enquiry into the Origin of Our Ideas of the Sublime and Beautiful* (1757) mentions the word "touch" thirty-three times. In the section "The Beautiful in Feeling," Burke also aligns sight with touch as Addison and More do: "Beauty, so far as it is taken in by the eye, may be greatly illustrated by describing the nature of objects, which produce a similar effect through the touch. This I call the beautiful in *feeling*. It corresponds wonderfully with what causes the same species of pleasure to the sight."[85] One description from this section in particular stands out in terms of corporeality: "All bodies that are pleasant to the touch, are so by the slightness of the resistance they make. Resistance is either to motion along the surface, or to the pressure of the parts on one another: if the former be slight, we call the body smooth; if the latter, soft. The chief pleasure we receive by feeling, is in the one or the other of these qualities; and if there be a combination of both, our pleasure is greatly increased." As I will discuss in the subsequent chapters, resistance to touch is paramount in terms of self-identity and recognition of the other (the beauty and pleasure Burke describes). Resistance confirms corporeality and tactility. It is this desire for confirmation of bodily existence and presence that informs the popularity of waxworks in the period (discussed in Chapter 3) and inspires tangible memorial objects.

Jonathan Reinarz and Leonard Schwarz underscore the significance of sensory history and the Enlightenment understanding that the senses inscribe the mind and "form the basis of our knowledge of the world."[86] Yet as Constance Classen observes in *The Deepest Sense: A Cultural History of Touch*, "Touch . . . is often downplayed or disregarded even within such fields as the history of the body or the history of medicine" (xi). She cites the centrality of touch in Daniel Defoe's narrative rendering of the Great Plague in *Journal of a Plague Year* (1722) to its marginalization in recent historical accounts. While sight dominated Western thinking during the eye-centered Renaissance and Enlightenment, recent studies have emerged that demonstrate that other senses, including touch, were also considered "authenticators of truth and couriers of reason" (Reinarz and Schwarz, 466). Addison's and More's descriptions attest to this intermingling of sight and touch, and the reliance on feeling and corporeality.

In an age of empiricism, we see this exploration of touch in numerous philosophical treatises. Though it is beyond the scope of this project to provide a complete analysis of Enlightenment sensory theory, to help support my claims it is necessary to examine a few influential theories of touch. The English philosopher John Locke in *An Essay Concerning Human Understanding* (1689) rejected the doctrine of innate ideas proposed by Descartes and the rationalists. In Book II he argues that every idea is derived from experience either by sensation—direct sensory information—or by reflection—"the perception of the operations of our own mind within us, as it is employed about the ideas it has got."[87] As an empiricist, Locke privileged the senses and offered an influential theory of touch. In his description of the spatial and temporal characteristics of touch, he emphasizes the cognition of solidity: "The idea of *solidity* we receive by our touch; and it arises from the resistance which we find in body to the entrance of any other body into the place it possesses, till it has left it. There is no idea which we receive more constantly from sensation than solidity."[88] For Locke, the primary qualities of an object are properties which the object possesses independent of us—such as occupying space, being either in motion or at rest, and having solidity and texture. Locke claimed that someone who lacks a sense will never be able to acquire the particular ideas pertaining to it (such as a blind man distinguishing color). Molyneux's Problem (or Question) signifies these conjectures and debates about the senses.[89] In 1688, the Irish scientist and politician William Molyneux sent a letter to Locke pondering the question of blindness and sensory capacities. To briefly summarize, Molyneux asked whether a man born blind who has learned to distinguish a globe and a cube by touch would be able to distinguish these same objects by sight if he was enabled, one day, to see. Though he did not answer the letter, years later Locke included the question in the second edition of his *Essay*, and it became a thought question for many Enlightenment philosophers (and as for Locke, he argued that it would take time for new sensory perceptions to develop and be able to distinguish and identity objects).

These notions of the primacy of solidity and touch are debated throughout the period.[90] Scottish philosopher David Hume opens *A Treatise of Human Nature* (1738) with "All the perceptions of the human mind resolve themselves into two distinct kinds, which I shall call IMPRESSIONS and IDEAS."[91] Hume takes "impressions" to be distinguished from ideas, on the basis of their force, liveliness, and vivacity. Ideas are therefore "faint" impressions. In a well-known tactile example, Hume explains that experiencing the painful *sensation of touching* the handle of a hot pan is more forceful than simply *thinking about touching* a hot pan. He explains in *Treatise*, "Thus we remember, to have seen that species of object we call flame, and to have felt that species of sensation we call heat." According to Hume, impressions are meant to be the original form of all our ideas. Sight and feeling (touch) are the senses from which we derive the first notions of space and extension. Hume states that the idea of space "is convey'd to the mind by two senses, sight and touch." The impression must resemble a substance, thereby Hume suggests that the ideas of sight and touch have a place and are extended, meaning that they have a shape. While saying that their extension is bodily or material can certainly be debated in Hume, "He says that perceptions have parts, and he applies the language of the body to them."[92]

Following Locke and Hume, Denis Diderot explored in *Letter on the Blind* (*Lettre sur les aveugles à l'usage de ceux qui voient*, 1749) man's reason and the knowledge acquired through perception. His letter, which recounted the life of Nicolas Saunderson, a blind mathematician, was intended to advance secular empiricism and disparage the religiously tinged rationalism put forward by René Descartes. The letter's discussion of sensory perception in men born blind dismissed the supposed primacy of visual imagery in abstract thinking. Diderot argued that touch is "the most profound and philosophical of the senses."[93] Burke's description also invokes the example of blindness and effectively compares touch and sight:

> The touch takes in the pleasure of softness, which is not primarily an object of sight; the sight, on the other hand, comprehends color, which can hardly he made perceptible to the touch: the touch, again, has the advantage in a new idea of pleasure resulting from a moderate degree of warmth; but the eye triumphs in the infinite extent and multiplicity of its objects. But there is such a similitude in the pleasures of these senses, that I am apt to fancy, if it were possible that one might discern color by feeling (as it is said some blind men have done) that the same colors, and the same disposition of coloring, which are found beautiful to the sight, would be found likewise most grateful to the touch.[94]

Burke goes on to compare textures and surmises "*smoothness* is a principal cause of pleasure to the touch, taste, smell, and hearing." (This particularity will be especially important in Chapter 3, in which I discuss wax and its smooth, fleshy texture.) Burke aligns this property with beauty, pronouncing, "I do not now recollect anything beautiful that is not smooth."

Following the work of Locke and subsequent empiricists, George Berkeley's *An Essay Toward a New Theory of Vision* (1709) theorized that the relation between visual and tactual sensations of objects differs. He believed that this relation is arbitrary and based only on experience. Ultimately rejecting Locke's system of primary and secondary qualities, as Paterson puts it he "argues for a negative answer to Molyneux's question . . . based on the radical incommensurability of sensations from sight and from touch."[95] For Berkeley, sensory experience and individuation are based on the theory that they are distinct, some senses being portals to spatial and haptic dimensions and others portals to nonspatial dimensions (vision).[96] He claims that "we never see and feel one and the same object" since they are perceived separately (*Vision* 45). Thereby, Berkeley expressed the idea that the visual qualities of a sphere are nonspatial and that a two-dimensional shape would not help the formerly blind to identify and distinguish shapes. He posited that a blind man restored to sight would not be able to distinguish objects and would have to learn to see.[97]

An adherent of Berkeley's theory of vision, Condillac also weighs in on this influential thought experiment. In *Treatise on Sensations* (*Traité des sensations*,

1754), Étienne Bonnot, Abbé de Condillac, adds that "touch, and touch only, instructs the other senses as to orientation, movement and activity," stating further that all of our knowledge comes from the senses.[98] For Condillac, touch then teaches us to attribute smells, sounds, tastes, and colors to external objects. In his well-known inanimate analogy, Condillac posits that when the "statue" touches another object, it experiences only one sensation of solidity without the answering sensation. The absence of the second sensation is supposed to induce the judgment that there is something else that is solid. An awareness of the shape of objects and their locations in an ambient space is then developed on this basis (*Treatise* III.i–ii). Touch is considered a sort of prelinguistic instrument of analysis: it originally discriminates parts outside of parts in its objects and by its means. Thereby, sensations become ideas, that is, they acquire a reference to external objects.[99] For Condillac (and later Kant), the mind must compare and judge manual experience in order to produce understanding.

In a similar fashion, Jean-Jacques Rousseau's *Émile, ou De l'éducation* (1762; English trans. 1763) argues for a return to direct sensory experience to determine truth. Rousseau writes, "Discrimination by touch is the surest just because of its limitations; for extending only as far as our hands can reach, it corrects the hasty judgments of the other senses, which pounce upon objects scarcely perceived, while what we learn by touch is learnt thoroughly."[100] Clearly, Rousseau favors active touch and recognizes its corrective power. Suzanne Clover Lindsay explains, "He made paramount the intimate experience of touch—exploring the physical world directly with the body's most extensive sense organ for attributes that confirmed its reality and complex identity. Rousseau and other sensualists rejected the received hierarchical division of the senses that favored sight (the faculty associated with global rational thinking); they considered it the least reliable and most easily deceived. Instead, many urged a renewed alliance of the faculties, with touch as their tutor, guide, and ultimate arbiter."[101] This collaboration of the senses, guided by touch, is relevant for a more urban, commodity-driven, manual labor–intensive culture.

Touch relies on flesh. We know we have touched another when their flesh resists. Flesh is the source of knowledge for the Enlightenment anatomist. In *The Senses of Touch: Haptics, Affects and Technologies*, philosopher Mark Paterson identifies that touch involves "the co-implication of one's own body and another's presence," the "co-implication of body, flesh and world."[102] *Death and the Body in the Eighteenth-Century Novel* works with Paterson's understandings of touch as "co-implication," "mutability of *flesh*," and "visceral" in order to expose the tactility of the body and the desire for its preservation in various modes. As

Newby and Toulson put it in their introduction to *The Materiality of Mourning: Cross-disciplinary Perspectives,* "Tangible remains play an important role in our relationships with the dead; they are pivotal to how we remember, mourn and grieve."[103] In subsequent chapters, I discuss the emotional reliance on the tactile and tangible in mourning practices in more detail.

In a period when empirical philosophers were theorizing humans' sentient knowledge, and scientists were penetrating, dissecting, and preserving the body as they pursued biological and physiological knowledge, it is no wonder that eighteenth-century art and literature explored the senses and corporeality, and that these have inspired numerous critical studies of sensibility and feeling, sentimentality, and vitality.[104] Roy Porter recognizes in the eighteenth century "a growing emphasis on the culture of the heart, on sensibility, and on private moral judgment."[105] To be sensitive and empathetic and to care for others (even fictional characters who are suffering) was tied to one's good moral character. Benjamin Barker-Benfield describes in his seminal study that the cult of sensibility first blossomed in the mid-1740s and made a powerful impact on European social and literary spheres.[106] The philosophical origins of this culture are well-documented, but I would like to underscore that the cultural ideals of restraint in the 1720s and 1730s engendered a violent reaction. As the role of external sense organs in mediating the mind/body relation was an ongoing debate, Anne C. Vila remarks that sensibility "provided both a dynamic vision of the body's interior, and an equally dynamic way of theorizing the interface between the inner and outer worlds of the human being—the liminal space where the external senses operated."[107] Barker-Benfield and Vila both cite Newton's widely read *Opticks* (1704) as an influence on the rise of sensibility. Newton speculated that light rays caused vibrations that traveled though the nerves into the brain and created the sensation of seeing, and this reframed "seeing" as a kind of "touching."[108] Bryson observes, "Touch was the dominant sense and metaphor of sensibility, for it involved both notions of symptomatic human connection, and the sexual frisson that secretly ran through this ostensibly moral cult."[109] Tracing the origins of human nature to the material and mechanical operations of the body's nerves was integral to thinkers like George Cheyne, Robert Whytt, and David Hartley. In *Observations on Man, his Frame, his Duty, and his Expectations* (1749), Hartley uses the terms "Touch" and "Feeling" interchangeably and privileges the sense of touch above others.[110] Feeling thereby becomes the new privileged mode since nervous operations are properties both within and between bodies. With these scientific accounts circulating midcentury, nerves, touch, and feeling materialize in literature, especially novels about men of feeling. As science pursued the connection between physiology and sympathetic feeling, novelists employed vocabulary that reflected sympathetic sensitivity, physicality, and overactive nervous systems.

Sensibility also became a "path of liberation"; the cult of feeling "was an escape from the repressive clutches of pagan stoicism" and offered a space to challenge gender norms (Craske 313). The body and how it is read is central to this experience of sensibility. John Mullan notes that "tears, blushes and sighs—and a range of postures and gestures—reveal conditions of feeling which can connote exceptional virtue or allow for intensified forms of communication."[111] The physiology of the sentimental body was thus highly gendered as women were supposed to have finer nervous capacities. Yet defining women as beings endowed with superior sensibility and heightened capacities for sympathy could also be restricting (and disempowering). The position of the sympathetic mourner was mostly a passive one. However, "men of feeling" emerge as figures that destabilize this emotional and physical stereotype. "Women, supposed irrational and emotional, were no longer 'other' to masculine experience, for their 'frailties' were shared by men."[112] Matthew Craske offers a helpful explanation: "Although the idea of the rational superiority of men was not forgotten, the possibility that men and women were formed by God and Nature equally passionate was considered."[113]

Mackenzie's Harley in the sentimental novel *The Man of Feeling* (1771) is exemplary in terms of male approaches to death and mourning. Hearing that Sir Harry Benson is about to marry his love Miss Walton, Harley is heartbroken. When he discovers the report is false, he still remains silent, not divulging his feelings to Miss Walton. After nursing the ill Old Edwards, and weary of the world's pain, he falls ill himself and gently retreats into death, which is seen as a gallant act. When Miss Walton visits him on his deathbed, and acknowledges his "worth" and her "esteem" for him, Harley experiences a shock of joy that is too much: "I feel something particularly solemn in the acknowledgment, yet my heart swells to make it, awed as it is by a sense of my presumption, by a sense of your perfections," he remarks.[114] When he seizes her hand, Harley falls dead: "He sighed and fell back on his seat—Miss Walton screamed at the sight—His aunt and the servants rushed into the room—They found them lying motionless together. His physician happened to call at that instant. Every art was tried to recover them—With Miss Walton they succeeded—But Harley was gone forever" (92). The violent conjoining of hands ("he seized her hand") is what seems to produce the shock to his nervous system. Human hands are intimate, visceral objects that physically affirm one's identity and vitality. While this touching exchange results in Harley's death and Miss Walton's deathlike collapse, a subsequent scene underscores the significance of touch as a confirmation of life. On visiting Harley's corpse, the male narrator describes his haptic reaction: "I entered the room where his body lay; I approached it with reverence, not fear: I looked; the recollection of the past crowded upon me. I saw the form which, but a little before, was animated with a soul which did honour to humanity, stretched without sense or feeling before me. 'Tis a connection we cannot easily forget:—I took his hand in mine; I repeated his name involuntarily; I felt a pulse in every vein at the sound. I looked seriously into his face; his eye was closed, his lip pale and motionless. There is an enthusiasm in sorrow that forgets impossibility; I wondered that it was so" (93). In this intimate scene with the corpse, the narrator's senses fool him while he touches the dead hand. He feels Harley's veins pulsing and fantasizes that he is still alive. Only the vitality of touch allows for this visceral response.

As the narrator describes, his interaction with Harley's corpse is one of "reverence, not fear." The corpse is a tangible, even lively object. In Mackenzie's depiction, death is an object of wanly, pleasurable contemplation for those, like Harley, who long to escape the world. With this, he is the male counterpart to Clarissa, a pure sentimental hero who embraces death. Violating sexual stereotypes, the sentimental hero, with his proneness to tears and effeminacy, is repeatedly held up by authors as reflecting natural and becoming traits, while for many, the woman of feeling tended to reinforce rather than subvert gender stereotypes. Parodied and maligned by the end of the century, but still relevant in the Romantic period, the cult of sensibility and the literature it produced are directly related to sensory investigations, religion, and, in part, approaches to death and mourning.

As we see a rise in the culture of sensibility from the midcentury onward, fashionable diseases and feeling men emerge as products of a feminized nervous system. Lawlor outlines this as such: "At the end of the seventeenth century new modes of medical thinking came into being, notably the partial displacement of the ancient humoural theory by a mechanistic Newtonian model and then, around mid-century, an increasingly dominant physiology of the 'nerves.'"[115] This medicine of the nerves, the solid tubes and fibers that carried sensations throughout the body, linked the sciences of physiology with the mind and "led to a paradigm shift in which both medicine and literature were dominated by notions of nervous sensibility" (49). A progression of nervous disorder to consumptive illness was key in many of the period's treatises, including those of George Cheyne, Robert Whytt, and Albrecht von Haller.[116] In Lawlor's reading, consumption becomes a fashionable disease for men as it signals "refinement"; consumptive males were thought to be more creative, intelligent, and poetic. For women, they became the frail, suffering, beautiful image of the creative male. Cheyne defines the resulting psychological disorders as the new "English Malady" (including the idea that the English could think of themselves as prone to suicide both in their literature and in the way they lived) and connects problems of the mind with further physical deterioration.[117] Cheyne posited that sedentary lifestyles of the wealthy, rich foods, caffeinated drinks, and alcohol all contribute to a weakening of the system. The middle and upper classes of a prosperous Britain were thereby especially "smitten with this disease of indulgence" (Lawlor 46). As charted in the introduction, by midcentury, there exists a thriving economy of emotive sensations to include pleasurable pain, as Burke outlines, and the appeal of melancholy. Thereby, sensibility is partly rooted in morbidity.

It is important to underscore that sensibility coincides with the immense popularity of novels, especially epistolary novels that purported to be "real" collections of letters between lovers and friends. Meant to instruct their readers on the etiquette of letter writing, social proprieties, and virtuous behavior, epistolary novels incorporated scenes of high emotion and sexual dangers, using "touching" (or arousing) scenes and sensory words (famously, Fielding's *Shamela* and Sterne's *Tristram Shandy* will parody these qualities). The cult of sensibility emerges in the eighteenth-century sentimental and "man of feeling" novels, such as Samuel Richardson's *Pamela* (1740), Laurence Sterne's *Tristram Shandy* (1759–1767) and *A Sentimental Journey* (1768), Oliver Goldsmith's *Vicar of Wakefield* (1766), Henry Mackenzie's *The Man of Feeling* (1771), and continental examples such as Jean-Jacques Rousseau's *Julie, or the New Heloise* (*Julie, ou la nouvelle Héloïse*) (1761) and Goethe's *The Sorrows of Young Werther* (*Die Leiden des jungen Werther*) (1774), which all present characters who possess delicate sensations, are easily moved to sympathy, or emotionally react to landscapes, art, and music.[118] Finding its origins in amatory fiction of the late seventeenth and early eighteenth centuries, the sentimental novel (or novel of sensibility) displayed the power of feelings as a guide to virtuous conduct (and was in part a reaction against religious doctrines that viewed humanity as inherently sinful and depraved). The immense popularity of epistolary novels reflected and generated "new medical-moral theories around sensitivity as an indicator of moral refinement and virtue" (Bryson 168). In its confirmation of the fashionability of female physical fragility and melancholy, the sentimental novel also engendered beautifully suffering heroines such as Clarissa, who takes control over her own death narrative. Thus, these popular narratives that present a model of tragic sensibility verge on the precipice of self-destructive affectivity.

For my discussion, the cult of sensibility's emotional and psychological significance, conjoined with its reliance on the body and its physical reaction to pain and beauty, is one of the many contributing factors to a growing sentimentality in mourning practices that cherished and physically reliquized the lost love one. And as the material afterlives of novels such as *A Sentimental Journey* and *The Sorrows of Young Werther* attest, sensibility and commerce were intertwined in the second half of the century. These two fictional examples demonstrate the resulting commodification of feeling and sympathy: a more heavily inflected consumer culture and the celebration of sensibility produced a traffic in things (snuffboxes, coins, dishes, and so on), a seemingly endless list of marketing products, including mourning objects. As the opening of this chapter points to, genuineness of feeling seems tethered to commodity exchange and circulation. Funerary practice is a profound forum in which to monitor these cultural movements, and in my reading, the novel in particular provides a broad expanse to work through and navigate changing attitudes toward death, mourning, and the body.

CHAPTER 2

Preserving Clarissa

Eighteenth-Century Embalming Practices

> A good book is the precious life-blood of a master spirit, embalmed and treasured up on purpose to a life beyond life.
>
> —John Milton, *Areopagitica*

> A book is a dead man, a sort of mummy, embowelled and embalmed, but that once had flesh and motion, and a boundless variety of determinations and actions.
>
> —William Godwin, *Fleetwood*

Toward the close of Samuel Richardson's *Clarissa, or The History of a Young Lady* (1747–1748), a novel obsessed with corporeality, we find a sequence of correspondence between Lovelace and Belford that expresses not only Lovelace's morbidly obsessive desires, but the tradition of embalming within his own and, we can assume, other wealthy families. On learning of Clarissa's death, Lovelace proposes to Belford that a horrible dissection and parceling of Clarissa's corpse be performed for his own pleasure and benefit:

> I Think it absolutely right that my ever-dear and beloved Lady should be opened and embalmed. It must be done out of hand—this very afternoon. Your acquaintance Tomkins and old Anderson of this place, whom I will bring with me, shall be the Surgeons. I have talked to the latter about it. I will see every-thing done with that decorum which the case, and the sacred person of my Beloved require.

> Every-thing that can be done to preserve the Charmer from decay, shall also be done. And when she *will* descend to her original dust, or cannot be kept longer, I will then have her laid in my family-vault between my own Father and Mother. Myself, as I am in my *soul*, so in *person*, chief mourner. But her *heart*, to which I have such unquestionable pretensions, in which once I had so large a share, and which I will prize above my own, I *will* have. I will keep it in spirits. It shall never be out of my sight. And all the charges of *sepulture* too shall be mine.[1]

Lovelace, in this delirious fantasy of macabre possession, plans to have Clarissa's viscera buried separately from her body: "Her bowels, if her friends are very solicitous about them, and very humble and sorrowful (and none have they of their own), shall be sent down to them—To be laid with *her* ancestors—unless she has ordered otherwise" (8:45, letter 17, Mr. Lovelace to John Belford). He then sums up his requests in a list of demands:

> 1. To forbid you intermeddling with any-thing relating to her. To forbid Morden intermeddling also. If I remember right, he has threatened me, and cursed me, and used me ill—And let him be gone from her, if he would avoid my resentments.
> 2. To send me a lock of her hair instantly by the bearer.
> 3. To engage Tomkins to have every-thing ready for the opening and embalming. I shall bring Anderson with me.
> 4. To get her Will and every-thing ready for my perusal and consideration.

His last request is indeed disturbing: "I will have possession of her dear heart this very night; and let Tomkins provide a proper receptacle and spirits, till I can get a golden one made for it" (8:46, letter 17, Mr. Lovelace to John Belford). Though never enacted, and later denied by him, Lovelace's fantasies about preserving Clarissa's corpse suggest not only the morbid desire to "possess" her postmortem body (an attempt to keep her, in part, "alive," to own and control the corporeal), but also an anxiety about her bodily dissolution.[2] Lovelace's wish to preserve, display, and touch the dead body is not grounded in the remnants of Catholic reliquary tradition; in my reading, these moments reflect a growing sentimentalism toward what I term "secular relics."[3] Clarissa's preserved remains are to be treated as icons of her enduring purity, as well as reminders of Lovelace's sin, trauma, violence, and desire. And as if prophesying that her corpse will be somehow fetishized by Lovelace, Clarissa directs in her will that she "may be put into my coffin as soon as possible" and interred in the churchyard "between the hours of eleven and twelve at night" (8:98, letter 27). In addition, she requests that he not be able to view her body:

> It is my desire that I may not be unnecessarily exposed to the view to any body; except any of my relations should vouchsafe, for the last time, to look upon me.
>
> And I could wish, if it might be avoided without making ill-will between Mr. Lovelace and my Executor, that the former might not be permitted to see my corpse. (8:98, letter 27)

Though somewhat shocking to a modern reader, possibly the reason that this particular correspondence in the novel has received little critical attention, this scene of virtual embalming, of macabre *memento mori*, in fact reflects, though in an extreme manner, the sensibility of the time, as I have argued elsewhere.[4] Yet Richardson's inclusion of Lovelace's preservation fantasy prompts an important question: *Were* people embalmed in the eighteenth century?

Previous to this study, the only published work devoted to eighteenth-century embalming practices, Jessie Dobson's pioneering "Some Eighteenth Century Experiments in Embalming" (1953), provided a detailed overview of the Hunter embalming methods along with descriptions of the outlandish "curiosities" housed by the Hunter brothers and others who were experimenting with embalming.[5] Dobson does not provide information regarding the commonality or reception of modernized embalming techniques and practices, nor does her study address the cultural context of embalming. Julian Litten's chapter "Taintless and Pure: Embalming Techniques" in *The English Way of Death* (1991) borrows from Dobson's essay, provides additional evidence of the practice of embalming, but relegates it to extreme cases. This now leads us to ask another question: How often was embalming practiced in eighteenth-century Britain?

Many critics and historians firmly insist that embalming was practiced, though they provide scant documentary evidence. In *The Female Thermometer: Eighteenth-Century Culture and the Invention of the Uncanny* (1995), Terry Castle argues that in the eighteenth century, "the arts of embalming and even mummification (one thinks of Bentham's corpse) became common practices among all but the very lowest classes."[6] Robert A. Aubin claims that by the early eighteenth century, "embalmment was rapidly becoming *de rigueur*. It paraded commercial and social respectability, followed Roman example, and was in line with the sentimental necrophily of the time."[7] In *Death, Burial and the Individual in Early Modern England* (1984), Clare Gittings asserts that by the eighteenth century, embalming was practiced by all except the lower classes and that "embalming had been divorced from its original raison d'être—to allow time for the preparation of an elaborate or distant funeral—and had become simply a matter of social prestige and snobbery" (92). This class distinction is evident from the funeral bills and costs. In 1694, the apothecary for Queen Mary II was paid £200 for performing the "customary funeral rites," and in 1760, the sum of £432 13s was received by the royal surgeons and apothecaries for the embalming of George II (Gittings, 228–29). Based on his study of burial registers, Stephen Porter contends that "the proportion of burials which involved a longer delay—even in some cases a far longer one—suggests the comparatively common use of embalming, if only among the wealthier sections of seventeenth-century society."[8] He states, "A wealthy person was more likely to have an elaborate funeral, with the corpse, embalmed if necessary, deposited in a leaden coffin which could be stored either in the house or in the church itself prior to burial."[9] In the register of St. Mary Woolnoth, Porter finds the longest delay between death and funeral to be forty-four days, an interval attributed to the socioeconomic standing of the deceased (Sir George Viner) and the elaborate preparations for the funeral. In *Death, Disease and Famine in Pre-Industrial England* (1976), Leslie Clarkson argues that in the seventeenth century, "among the middle and upper classes it was a fairly common practice to embalm the body" (153). Clarkson continues: "Embalming may have fallen a little out of fashion in some circles in the early seventeenth century, but it was still common among the well-to-do a century later. One gentleman dying in 1691 willed fifty guineas to his doctors and apothecaries for embalming him, another fifty for his coffin."[10] Though some would dispute these claims, I have found that church excavations, diaries, and wills indicate the popularity of incorporating various forms of embalming methods into burial rites.[11] Despite the fact that we are not certain about the percentage of bodies embalmed, literary examples, the consistent publication of treatises, lectures, and sermons, as well as the frequent circulation and display of preserved remains provide indisputable evidence that embalming and its history were certainly a part of eighteenth-century cultural interest.

As this chapter will detail, not uncommon, embalming in the eighteenth century helped shift the overseeing of death from the church to the individual and family, led to the rise of the undertaking trade, was often a mark of social class, and witnessed innovations that revolutionized anatomical study and surgical practices. And as discussed in Chapter 1, experience of and desire for solidity, tactility, and corporeality are bound in fiction that represents and cherishes the dead and dead-like body. Preservation of the body and the collecting, ownership, and circulation of its parts also morbidly reflects the empirical drive of the period. Battling decomposition, anatomists created wet and dry specimens, wax-forms, and atlases of the dissected body in order to preserve the knowledge they cut into flesh. Fearing dissection by the anatomist, parceling by the undertaker, or bodily dissolution and putrefaction, many sought aesthetic ways to immortalize the dead body. Highly personalized funerals, planned by the deceased themselves and rejecting established rituals for a more individual approach, occurred in the period throughout England. In *Death, Memory and Material Culture* (2001), Elizabeth Hallam and Jenny Hockey recognize this growing need for preservation in the period: "The pollution and contamination that the dead body therefore signifies is contained and managed in medicalized settings and through expert practices that are seen to protect public health. The corpse is regarded as a material manifestation of death, a body devoid of a self and individuality, but for the short time allowed for bereaved relatives to remain in contact with it, the body is 'staged' and presented in an approximation of embodied 'life.' This is attempted through embalming, restoration, the application of cosmetics and the use of clothing that temporarily arrest decomposition."[12] While Hallam and Hockey identify these practices as temporary, and vital to the visual workings of memory, I recognize a more tangible process, one that allows for permanent or semipermanent preservation. Advanced embalming techniques could indeed better forestall fears as well as decay. Brian Parsons recognizes that "Embalming was practised in England during the eighteenth century,"[13] and as Ruth Richardson puts it in *Death, Dissection, and the Destitute* (2000), "embalming was in the news at the time" (171).

It can be argued that embalming was made necessary in preindustrial England because of the "length of time that often elapsed between death and burial, an interval during which relations and friends were summoned and preparations made for the funeral" (Clarkson 153). Historians agree that embalming was absolutely essential if a grand burial, such as the heraldic funeral of a member of the aristocracy, was planned (Gittings 104). Prior to the eighteenth century, the process included eviscerating the corpse, immersing the body in alcohol, inserting preservative herbs into incisions previously made in the body, and wrapping the body in tarred or waxed sheets.[14] The British Institute of Embalmers acknowledges that Cnut II (1040–1042), the Danish king of England, was embalmed by these or similar methods, as were William the Conqueror (1066–1087), Henry I (1100–1135), and Edward I (1272–1307). Unfortunately, the revolting result of embalming Henry I (it was known that his corpse "exploded") provides an example of how primitive preservation techniques could fail. King Edward's tomb in Westminster Abbey was opened in 1774 and a description of the blackish-brown desiccated corpse—still in excellent condition almost 500 years later—was recorded. Cnut's body was still in a state of good preservation when it was discovered in Winchester Cathedral in 1776. And William's body was found well-preserved in France in the sixteenth century. Samuel Pepys's infamous entry from February 23, 1669, on kissing Queen Katherine of Valois (d. 1437) on his birthday ("I did kiss a queen") attests to the fact that embalming combined with proper coffining could work amazingly well.[15] In the 1680s, apothecary Thomas Warren placed advertisements in *The London Gazette* for his "most curious and excellent way of preserving Dead Bodies from Putrefaction, change of Colour, or Complexion, without Disbowelling, Sear-cloathing, Mangling or Cutting any part thereof, to the great Approbation of several of His Majesties Physicians."[16] Charles II quickly secured him as "His Majesties Apothecary in Ordinary." Richard Mead (1673–1754) and the botanist Sir Hans Sloane (1660–1753) experimented with and practiced embalming, and Dr. Mead eviscerated and embalmed the body of Queen Anne (d. 1714), "whose method, it was declared, omitted nothing for its preservation."[17] The drawn plans for the King Henry VII Chapel in Westminster Abbey (c. 1737) show placement of numerous monarchical viscera urns (fig. 2.1). Notable is the description of Prince George's funerary arrangements (he died in 1718); of those listed, the only to include reference to "bowels": "D.D. The Urn which Contain the Bowells of Prince George." Objected to by Queen Elizabeth (d. 1603) and Queen Mary II (d. 1694), the embalming of British monarchs in fact continued without disruption until the reign of Victoria.[18] For the monarchy, physically preserving the body implies the continuity of political and divine authority, making the corpse a symbol of posthumous power, of long-term political and social effect. It is no surprise, then, that the aristocracy took to aping these burial and mourning practices, which in turn led not only to the commodification of death in the eighteenth century, but also to the rise of the undertaker's "art" of embalming. Craske remarks that "Undertaking was perceived to cater to the pretentious nouveaux riches of the City, providing heraldic ceremony to those without entitlement" (45).

For this investigation, it was crucial to find instances of the practice outside of monarchical tradition. Though no quantitative analysis has been completed, we can look at church reorderings for evidence that embalming was consistently performed. The overcrowding of churchyards often resulted in the reorganization of mausoleums and graveyards. These reorderings suggest that many of the

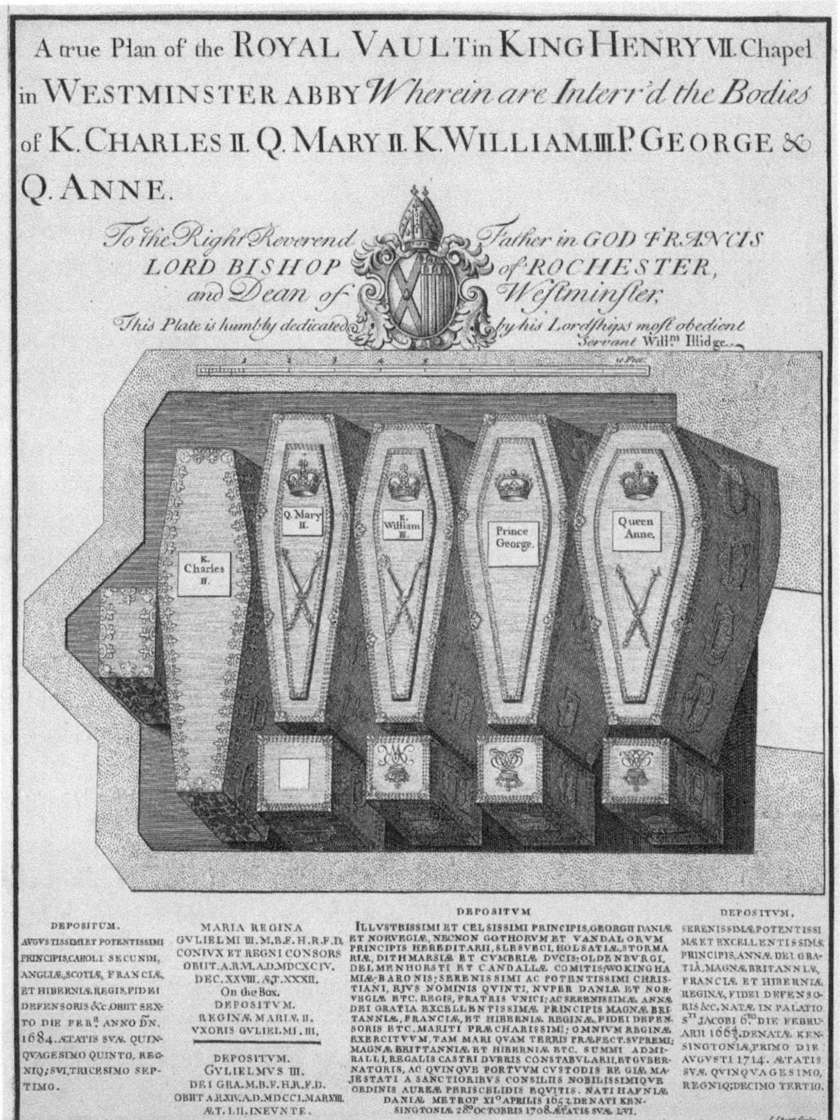

Figure 2.1. Prince George urn, King Henry VII Chapel etching, Westminster Abbey, c. 1737. © The Trustees of the British Museum.

wealthy were indeed embalmed, and that often prearterial embalming practices were quite successful. Publicized accounts kept embalming in the social conscience and reiterated the acceptance and popularity of the art even before eighteenth-century innovations.[19] Unfortunately, organized reorderings and vault excavations are only recent phenomena. Previously, churches and undertakers did not keep close records of embalming instances, and reorderings did not often include a list of coffin contents or the presence of viscera boxes. This reluctance, or resistance, persists. Modern excavations provide detailed histories and descriptions of coffins, but patently exclude mentioning embalming evidence. Exhuming or moving coffins often involved the curious act of opening them and describing the state of their contents. In *Artifacts: How We Think and Write About Found Objects*, Crystal B. Lake describes four exhumations of English monarchs conducted between 1774 and 1813, and antiquarian interest in "these sovereign bodies as pure matter, even dead matter."[20] Most of the recorded incidents were published in periodicals such as *The Gentleman's Magazine* and in British histories such as Sir Thomas Browne's *Hydriotaphia* (1658) and William Burton's *Leicestershire* (1622). Several of the accounts of embalming describe aristocratic families of the fifteenth and sixteenth centuries. Charles Collignon, a professor of anatomy at Cambridge, read "Some Account of a Body lately found in uncommon Preservation, under the Ruins of the Abbey, at St. Edmund's-Bury, Suffolk" to the Royal Society on June 25, 1772.[21] Collignon explained that workmen digging among the ruins of the abbey discovered the leaden coffin of Thomas Beaufort, Duke of Exeter (d. December 31, 1426), uncle to King Henry V. Thomas Cullum, the physician at the scene, informed Collignon that the viscera had been removed, that the remaining body parts were amazingly entire, and that "the fluid running out of the coffin, upon its being moved, might occasion the suspicion of the body being put in pickle" (466). Noting that some bodies are less prone to putrefaction than others, Collignon states that if "we be masters of all these particulars . . . it would lead, perhaps to the probable cause of the phenomenon, and point out a proper method of imitation." He admits, "It is difficult to know how much merit is to be assigned to the art or mystery of embalming, and how much to the power of natural causes" (468). Collignon's account demonstrates the interest in improving embalming practices and methods, and explicitly endorses the benefits of preservation. It also reflects the same difficulties that exist today in the field of cemetery archaeology.

Julian Litten notes that during the 1984 reordering of St. Mary's, Little Ilford, in Essex, the lead coffins of John (d. 1737) and Charles Lethieullier (d. 1759), the sons of famed antiquarian Smart Lethieullier, were examined. Charles's coffin was filled with a honey-like liquid. The viscera chest was opened and found to contain the "soft organs and heart, still elastic and resistant to the touch."[22] Despite Litten's observations, the church website and excavation papers exclude any mention of the Lethieullier family's history of embalming. In the case of the excavation of the London vaults at Christ Church Spitalfields from 1984 to 1986, Margaret Cox recorded the archaeological findings of approximately one thousand coffins of mainly middle- to upper-middle-class occupants. In *Life and Death in Spitalfields: 1700 to 1850* (1996), she remarks, "The state of preservation of the human material varied from virtually complete, including skin, hair and internal organs, to a sediment of crystal debris being all that remained of the bone."[23] She surmises that "particular conditions within some of the coffins led to mummification of the remains," due possibly to triple coffining and lead coffins, but admits that her excavation could not determine if the corpses had, in fact, been embalmed. Only one viscera box was found, and it did not contain human remains, but this leads Cox to say, "The presence of a viscera box in one coffin suggests that some embalming did take place" (111). The state of many of the remains made testing unlikely, and Cox's team did not test for the presence of historical embalming fluids even on those remains that might have been suitable. Either way, it would be impossible to determine how many of the remains did receive some form of preservation, which is another example of the difficulty of finding scientific evidence in this particular field. As the burial register revealed, the average interval between death and burial was found to be seven days, the longest being twenty-one days. Therefore, it is highly probable that many of the deceased did undergo some form of preservation. As the Spitalfields case suggests, similar studies of eighteenth-century cemeteries and burial registers are desperately needed in order to provide a fuller picture of the frequency of burial delays, as well as the probability of postmortem preservation.

Preserving a Good Death

While excavations continue to reveal the evidence of embalming practices, we must look at a culture's beliefs to help explain how a fascination for preservation could persist. As *Death and the Body in the Eighteenth-Century Novel* maintains, the most important sign of shifting attitudes in the eighteenth century is the emergence of a sentimental attitude toward death and the body. Beautification was used to hide the physical signs of mortality and decay and to overcome any sense of separation or loss of individuation. According to Gittings, since the

Renaissance, "unease at the prospect of physical decomposition led to the widespread coffining of bodies, and encouraged the craft of the embalmer" (13). And while many churches approved embalming for their clergy and aristocratic members, and the Catholic Church still embalmed its popes, in post-Reformation England funeral rites changed, and as a desire to observe the body of a much-revered individual in the home became more accepted, the art of the embalmer was more in demand.[24] The art was no longer reserved for the monarchy or associated with popery, and its religious purposes were both metaphorical and actual. And as regulation and control by the church and state over burial practices weakened in the period, personalized funerary and mourning practices emerged.

An unusual illustration of changing ideas about the corpse can be found in the account of Harriet Spencer (1761–1821), the sister of the Duchess of Devonshire, who recorded an occasion when her father, Lord Spencer, took her to see some mummified corpses in a church crypt, because "it is foolish and superstitious to be afraid of seeing dead bodies."[25] This unease in some degree stems from inspirational treatises on how to die a "good death" and sermons that emphasize the body's putrefaction and the immortality of the soul. Reverend Thomas Doolittle, in the Puritan sermon *A Serious Persuasive to Prepare for Death* (1755), reminds his listeners that death is "the destroying and demolishing of the Body of Man, that famous and curious Fabrick, and a bringing it into Dust and Putrefaction, (Psalm 40:3). It turns a living Body into a dead Carcass, a lifeless Lump of Clay; and causeth it to become Meat for Worms to feed on, (Job 18:26)."[26] John Bowden's sermon *The grave, the house appointed for all living. A sermon preach'd at the funeral of Mr. James Blunt, of Frome on Thursday the 19th of August, 1749* uses the fear of the body's dissolution to inspire listeners to prepare for death.[27] Bowden says, "What a great and surprizing change this must be! . . . to become senseless as a Stone; deform'd and frightful to behold; unfit to stay about Ground; a Lump of Rottenness, the Spoil of the Grave, and Food for Worms." Inspirational books on "how to die well," such as John Kettlewell's *Death made Comfortable: or the Way to Die Well* (1702) and John Hayward's *The Horrors and Terrors of the Hour of Death and Day of Judgment* (1707), traditionally incited fear concerning the body's decay in order to turn heathens to the Christian religion.[28] Whether or not physical resurrection would actually take place, embalming served as a metaphor for the possibility of providing a secure future for the soul. For this reason, phrases such as "embalming" a name or "preserving" memory are persistently used in eighteenth-century funeral sermons.

Though Clarissa was never embalmed, in death she is aligned with the marker of sainthood: an uncorrupted corpse. On the body's arrival at her father's house, Colonel Morden (Clarissa's cousin and subsequent "funeral journalist") writes a detailed description of the "fair corpse" for Belford, Lovelace's friend, who does not attend the funeral: "The corpse was very little altered, notwithstanding the journey. The sweet smile remained" (8:74, letter 21, Colonel Morden to John Belford). Clarissa's family is amazed at the corpse's condition: "She was their very niece . . . The same benignity of countenance! The same sweet composure! The same natural dignity—*She* was questionless happy! That sweet smile betokened *her* being so" (8:74–75, letter 21, Colonel Morden to John Belford). The viewers claim that "they never saw death so lovely before" and that death had not altered her features. Her beauty in death reminds the viewers that Clarissa has transcended human sinfulness; her saintly preservation represents the "good death" to which most Christians aspired. The uncorrupted body thereby becomes a physical symbol of a moral and virtuous life. As discussed in more detail in Chapter 1, David Cressy and Ralph Houlbrooke both affirm that people still strove for a good death well into the eighteenth century;[29] a good death, including a beautiful and peaceful deathbed scene, exemplified a virtuous life.

The belief that the bodies of Christian saints and other venerated individuals are incorruptible is ancient; the preserved body was often seen as a sign that the individual was holy or was favored by God. The Church of England never took an official position toward embalming, but seventeenth- and eighteenth-century sermons and inspirational texts often focused attention on the state of the interred corpse, perhaps suggesting embalming as a means of maintaining a corpse fit for resurrection. (Cremation was not yet in Christian practice.)[30] Ralph Brownrig, in *Sixty Five Sermons* (1674), emphasizes treating the corpse as a possession of the Holy Ghost, stating, "The Temple of God must not be destroyed."[31] In *Everlasting Blessedness. A Sermon Preached at the Funeral of . . . Mr. William Baker* (1692), Edward Dent remarks that "The Bodies of Saints in the Grave . . . are in Union with the Lord Jesus. For . . . if the Soul only has Union with Christ, then is the Body lost."[32] Clergy frequently mentioned the Jewish tradition of embalming the corpse with "sweet odours" as well as "her who brake the Box of precious Ointment for his Burial."[33] In the early sermon "The Art of Embalming Dead Saints" (1641), preached at the funeral of Master William Crompton, the pastor of a church in Launceston, Cornwall, George Hughes describes how the bodies of saints have been traditionally embalmed in order to secure them from "Corruption in the grave." Using Psalm 16:10 as a basis for his sermon, "For thou wilt not leave my Soule in Hell: neither wilt thou suffer thine Holy one to see Corruption," Hughes, in fact, compliments Crompton by suggesting that he deserves to be compared with the holy saints and therefore should not suffer bodily corruption. Hughes argues that the saints are Christ's "sons" and by "gracious imputation" are free from "putrefaction and Rottennesse" (20). Despite the eulogy, Crompton was not embalmed, but this alignment of embalming with religious expression occurs throughout the eighteenth century, not only in sermons, but in medical treatises as well.

Undertaking Embalming: Hearts, Bowels, Urns, and Pyramids

It cannot be denied that preserving the heart in spirits, using viscera jars, and burying bodies in vaults were accepted funeral rituals among the wealthy. Many eighteenth-century wills mention the deceased's request for embalming. An early example is the lengthy will of draper and philanthropist Francis Bancroft (d. 1727), which was published in its entirety the year following his death. Specifying his own funeral and burial arrangements, Bancroft writes, "My Body I desire may be embalmed within six Days after my Death, and my Entrails to be put in a Leaden Box, and included in my Coffin, or placed in my Vault next the same, as shall be most convenient."[34] He directs that he be buried in a vault within "ten Days after my decease" (6) and that his funeral "shall not exceed the Sum of Two Hundred Pounds" (7). Bancroft's will demonstrates that the management of death shifts from the church to the individual and professional undertakers. Excerpts of the draper's will were published several years later in the *Ladies Magazine*, which comments on its remarkable detail and seems to support Bancroft's planning and meticulous orders for his posthumous endurance.[35] According to Matthew Craske, surgeon Francis Douce (d. 1760), who also practiced embalming, goes to the unusual lengths in his will of requesting a particular undertaker to handle his corpse: "Mr. Hazaldine at the Mitre in Pudding Lane should be employed and, should he be dead, Mr. Keele at Upholsterer's Hall in Leadenhall Street."[36] In an eccentric nod to the original embalmers, Douce and his architect John Blake had a pyramid-style vault built in 1748 in Lower Wallop, Hampshire (fig. 2.2).

While others built Romanesque and Egyptian-style burial vaults, Douce's monument is unique in that it is also a tomb that contains internal burial rooms holding the bodies of both Douce and his wife.[37] While Douce built a functioning Egyptian-style mausoleum, most pyramidal structures were built as follies for design impact.[38] The Earl of Carlisle had a pyramid built in 1728 at Castle

Figure 2.2. Francis Douce mausoleum, Lower Wallop, Hampshire, c. 1748. Photograph courtesy of The Mausolea and Monuments Trust.

Howard to commemorate his ancestors, a pyramid was built by Vanbrugh in Stowe School (c. 1725) that is now destroyed, and Knill's Steeple in St. Ives, Cornwall, is home to a pyramid mausoleum (c. 1782, unoccupied) for John Knill, a former mayor of St. Ives. John "Mad Jack" Fuller (d. 1834), a Sussex eccentric, built a folly mausoleum, The Pyramid (c. 1811), under which he was buried and which still stands in the Brightling Churchyard next to his former home. James Burton (d. 1837) and other members of the Burton family are buried beneath the Burton Pyramid (c. 1837), St. Leonards, East Sussex. The Egyptian Revival–style Kilmorey Mausoleum (c. 1850) in St. Margarets, Richmond upon Thames, contains the bodies of the 2nd Earl of Kilmorey (d. 1880) and his mistress. The rise of the mausoleum is just one of many examples that reflect private control over funerary rites for the wealthy.[39]

Several advertisements in *The London Gazette* reflect embalming's growing acceptance in burial practices. An advertisement from December 24, 1684, announces William Russel's undertaking services:

> William Russel Coffin-maker, who hath the Art of Preserving Dead Bodies without Embowelling, Sear-clothing, Cutting, or Mangling any part thereof, and hath used it to the great Satisfaction of all those Honourable Persons by whom he hath been Employed, lives at the Sign of the Four Coffins in Fleet-Street. Coffins ready made and the Body Preserved for Five Pounds.

For this minimal cost, the process would most likely involve pickling. An instance from 1682 describes a strange case of an abandoned, embalmed infant, and the advertisement for £20 reward for further information:

> There having been a dead Female-Infant, of about a Fortnight old, Embalmed and wrapt up in Leather, lined with Melilot, enclosed in a well-made Leaden Coffin, and sent down in a silk Rug, fit for a Cradle, with other necessaries for a Child, within a Firr-Box, wherein it is supposed to have lain about two Months, left by an unknown Person at the Kings-Arms-Inn on Holbourn-bridge, to be sent to a Reverend Divine at Salisbury."[40]

Embalming and lead coffins are expensive practices, and as the description details, the parents of the child must have had some amount of wealth. *The London Gazette* of May 14, 1751, contains an advertisement for another "found" embalmed body: the commander of a Danish ship about six leagues off the Kentish Coast "took up a Coffin, made in the English Manner, and having the following Inscription engraved upon a silver'd Plate, viz. 'Mr. Francis Humphry Merrydith Died March 25, 1751. Aged 51.' That having carried the said Coffin (being of Oak, cover'd with black Cloth, and having Eight silver'd Handles,) to Hamburgh, and open'd it there, a Coffin of Lead was inclosed in it, which being also open'd, contained the Body of an eldery Man, embalmed, and dress'd in fine Linnen."[41] The Captain solders the coffin, preserving Mr. Merrydith on his ship "in a careful Manner," and agrees that if no one claims the deceased, he will return Merrydith to England for a proper burial at his own expense. These two notices are strange indeed, but as we will later find, they reflect and reinforce the practice of preservation for nonroyals and aristocrats.

Embalming was such an accepted funeral rite of the wealthy that it became an integral part of popular satires and visual culture. In Hogarth's *Gin Lane* (c. 1751), an undertaker's shop feeds off a declining society. Armed with coffins, and with predatory looks on their faces, undertakers search for bodies. Craske notes that undertakers were also known as "cold cooks" "on account of their preservation of corpses by soaking them in distilled spirits. This notion of the final pickling may well have been another reason for including them on Gin Lane" (40) (fig. 2.3). Early undertakers did practice preservation in "spirituous liquors." Craske refers to the 1728 corpse of Mr. Hillersden of Esltrow, preserved in this way, that burst into flames, causing a fire that burned down a row of London

Figure 2.3. William Hogarth, *Gin Lane*, c. 1751, Etching and engraving. 357 × 305 mm. Public domain.

houses (*Daily Journal*, October 7, 1728). The will of a wealthy London undertaker, John Purdey (PCC 1759, 105), leaves "brewing and distilling equipment" (Craske, 440n5). *The Death, Dissection, Will, and Funeral Procession of Mrs. Regency* (1789) is a humorous response to the withdrawal of the Regency bill following the recovery of King George from mental illness. The satire provides "an Account of the Opening of the Body" with amusing detail: "When the *sternum*, or chest was opened, all appeared right there; and her *heart* was sound and strong. The *bowels*, contrary to the idea of the surgeons, corresponded with the *head*, and were *faultless*."[42] Perhaps in homage to Richardson, the anonymous author then describes "the Emblematical Ornaments" on the coffin, including the drawing of "Virtue in Prince Royal robes" (11). The *Admirable Satire on the Death, Dissection, Funeral Procession, & Epitaph, of Mr. Pitt* (1795) is a more personal jab; it includes a lengthy description of Pitt's deathbed blunders and eventual death, followed by the opening of the dead body: "On sawing through the *cranium*, the first thing that struck an observer, was a remarkable accumulation of the *brain* on the *left* side of the skull, while the cavity on the *right* side was almost empty" (11). The heart is described as "frozen into a solid lump" (12). These satiric pamphlets demonstrate the extent to which embalming and its expected role in the funeral rituals of the wealthy penetrated the cultural landscape.

It should be no surprise, then, that Lovelace would follow burial practices that were not only traditional, but expected of wealthy families. This would include some form of preservation, as well as viscera burial.[43] "Her bowels, if her friends are very solicitous about them, and very humble and sorrowful (and none have they of their own), shall be sent down to them—To be laid with *her* ancestors," he states (8:45, letter 17, Mr. Lovelace to John Belford). Lovelace's request that Clarissa's bowels be buried separately would, of course, require the use of an urn, a popular eighteenth-century motif. Though he never produced urns for viscera burial, "in the mid-1770s Josiah Wedgwood started producing ornamental urns in the form of copies of Greek vases, some of which appear to have been used as memorial vessels," such as the black mourning urn to Henry Earle from 1774.[44] One scene features a weeping angel and the other a woman in mourning at the gravesite. Most commonly the viscera were placed in plain jars or lead boxes on shelves within a family crypt. In *A glimpse at the monumental architecture and sculpture of Great Britain* [. . .] (1824), historian Matthew Holbeche Bloxam states that "It was a common practice, when the body was embalmed, to take out the heart and bowels, and inter them in a different church to that in which the body was buried; and a request that this might be done was sometimes inserted in the will of the deceased. From numerous instances it appears that this custom prevailed from the twelfth to the eighteenth century."[45] The second edition of William Salmon's *Synopsis Medicinæ, or a Compendium of Astrological, Galenical, and Chymical Physick* (1681) includes a chapter on embalming that describes this evisceration and urn practice: "Embowel the Corps, removing all the Viscera or contained parts in the three Venters, except the Heart and Kidnies, which may be Embalmed with the Body. All these Bowels you may put in an Urne or pot, and Bury them by themselves."[46] In *Practical Surgery* (1796), Robert White directs that "The brain and viscera well cleansed, and covered with the aromatic powder . . . [be] put into a leaden chest, and soldered down" (362). If a death occurred overseas or at sea, as in Merrydith's case, often the body would be embalmed and the viscera returned in a plain jar filled with wine. In December 1805, immediately following the Battle of Trafalgar, Lord Nelson wrote that he accompanied his brother's corpse to London, was planning a funeral for the month following, and remarked to the presiding reverend, "I could wish to have known what has been done with the bowels—whether they were thrown overboard, or whether they were preserved and put into the coffin with the body."[47] His concern demonstrates the social acceptance of viscera burial and its emotional role in burial practices. This practice continued throughout the Romantic period, a notable example being that of Princess Charlotte, who died in childbirth November 6, 1817. The writer in *La Belle assemblée, or, Bell's court and fashionable magazine* describes, "The heart and intestines of this illustrious lady were, after being embalmed, deposited in an urn . . . and the outside exhibited a splendid covering of crimson Genoa velvet."[48] So as bowels are a source of disgust, especially associated with the lower and criminal classes (notoriously depicted in Plate IV of Hogarth's *The Four Stages of Cruelty* [1751]), and part of the now disputed humoral theory of medicine, they also receive sentimental associations, especially for the upper and aristocratic classes.[49] This dual register of associations is a key component of my larger argument: the culture was renegotiating secular and religious notions of both the reviled and the revered body.

As seen with Princess Charlotte's funerary rites, Lovelace's additional desire that "I will have possession of her dear heart this very night; and let Tomkins provide a proper receptacle and spirits, till I can get a golden one made for it" (8:46, letter 17, Mr. Lovelace to John Belford) reflects not only a common ancient burial rite, but also a traditional sentiment of mourning. If an entire body was not embalmed, survivors would request a piece of the body. Ariès remarks that "the part most sought after, the noblest part, was the heart, the secret of life and emotion" (387) (and it is notable that the most common symbol in eighteenth-century love jewelry is the heart).[50] The practice of burying the heart apart from the corpse was common from the twelfth to the eighteenth century, particularly for the funerals of kings and warriors.[51] Indeed, in the Middle Ages, many followed the growing custom of burying the heart at a favorite church or shrine, or even requested heart burial in the Holy Land.[52] As the practice most likely began in France, it has always been most popular in that country, but there are many instances of British royals and religious figures having their hearts and viscera separately buried, including during the seventeenth and eighteenth centuries, including the Bishop of Winchester in 1618, Oliver Cromwell in 1658, William III in 1702, Queen Anne in 1714, and George II in 1760.[53] William, Duke of Gloucester (d. 1700), the son of Princess (later Queen) Anne, died at age eleven, was emboweled and embalmed, and was placed in a lead coffin.[54] The heart and viscera were enclosed in a lead urn, secured in a square box covered with black velvet, and placed in the Stuart vault at Westminster Abbey, according to royal tradition.

Aristocrats, poets, and military members also have a tradition of burying their hearts under monuments, in family vaults, or at other locations with personal significance. The diplomat Sir William Temple (d. 1699) was buried in Westminster Abbey next to his wife, but he had requested that his heart be buried in a silver box under a sundial in his garden. James Radcliffe, last Earl of Derwentwater (d. 1716), requested that his heart be enclosed in a casket and be sent to the Church of the English Benedictine Nuns at Pontoise (where many others had also sent their hearts). Charles Douglas, Earl of Selkirk (d. 1739), was opened and embalmed and laid in a lead coffin in a vault in Westminster Abbey; his heart resides in an urn nearby. St. Paul's, Hammersmith, contains Sir Nicholas Crisp's heart urn, with the inscription, "Within this urne is entombed the heart of Sir Nicholas Crispe Knight and Baronet. A loyall Sharer in the Suffrings of his late and present majesty. He first settled the trade of gould from guyny and there built the castell of cormantine died the 26 February 1665 aged 67 yeares" (fig. 2.4).[55] Crisp was buried at St. Mildred's, Bread Street, but he requested his heart be embalmed, placed in this urn, and mounted below the monument to Charles I. The poet and satirist Paul Whitehead ordered that his heart be wrapped in lead and enshrined in an urn, costing £50, with this inscription:

Paul Whitehead of Twickenham, Esq., ob. 1775.
Unhallowed hands, this urn forbear,
No gems, nor orient spoil
Lie hear concealed; but what's more rare,
A heart that knows no guile.[56]

Figure 2.4. Heart urn, Sir Nicholas Crisp (d. 1666).
Courtesy of St. Paul's, Hammersmith.

The ceremonial at the heart burial included grenadiers, flute players, bassoon players, and fifers.. Whitehead's urn was borne by six soldiers and placed on a pedestal in the Dashwood mausoleum in West Wycombe churchyard. Apparently, his heart was sometimes taken out to show to visitors, until 1829, when it was stolen.[57]

Often the ornate receptacle for the heart was made of gold, silver, or wood. Philbert Guybert's *The Charitable Physitian* (1639) explains that after a heart has been removed and embalmed, it should be put into "a case of Lead, Silver, or Pewter, fashioned in the forme of a Heart" in order that it may be carried "whither you please." In *Chirurgorum comes* (1696), surgeon Alexander Read instructs, "If it be the Friends pleasure either to put the Heart in its place again, or to keep it apart in a box, let it first of all be washed with *Aqua nostra Balsamica ex aloe & myrhha*, afterwards with *Spiritus Solomonis*, and then let it be anointed with *Balsamus Peruvianus*" (711–12). In his study *Practical Surgery* (1796), Robert White also recommends an ornate container: "The heart, after being properly cleansed, and its cavities well filled with the powder and sewed up, [is then] placed in a silver urn" (362). William Hogarth dramatizes the ritual in his painting *Sigismunda* (1759) from *The Decameron* (1349–1351), which illustrates the tragic figure embracing a golden cup containing the heart of her lover Guiscardo, who has been murdered by her father (fig. 2.5).[58] Notably, Sigismunda has removed the casket's top, exposing the glossy red heart, which she touches with her fingertip. (In the tale, she adds poison to the heart casket and commits suicide by drinking it). Heart burial occurs throughout the century and extends into the Roman-

Figure 2.5. William Hogarth, *Sigismunda Mourning over the Heart of Guiscardo*, 1759. *Oil on canvas*. 39 × 49½ in. Photo © Tate, London.

tic period, the most notorious examples being that of Lord Byron (who having died in 1824 in Missolonghi, Greece, was embalmed in spirits, his body shipped to England), Napoleon (who died in 1821), and Percy Bysshe Shelley (who drowned in Italy in 1822). Byron was buried in the ancestral vault of the Byrons in Nottinghamshire, and his viscera were embalmed in an urn, which bears the following inscription: "Within this urn are deposited the heart, brains, &c, of the deceased Lord Byron." Following an autopsy that determined Napoleon died from stomach cancer, his surgeon removed his heart, intending to comply with Napoleon's wish that it be sent to his wife, Marie Louise, in a silver vessel filled with wine.[59] Shelley's remains were buried in the Protestant Cemetery in Rome, but his heart was sent to his widow, Mary, who kept it in a silken shroud and is said to have carried it with her for years (some biographers claim she preserved it in a book of Keats's poems, others in the pages of *Adonais*, Shelley's elegy for Keats, or that it was wrapped in silk). Found in her desk, the relic was buried with her on her death in 1852.[60] As these sentimental examples speak to, heart burial embodies the extravagant emotional response to the loss of a loved one, thereby finding a significant place in the task of the undertaker as well as the embalming surgeon.

The Modernization of Embalming Techniques and Practices

Information that appeared in frequently published medical books and treatises confirms what we learn from examining urns, heart burial, and other forms of preserved remains, namely that embalming was consistently practiced and discussed. In fantasizing about Clarissa's funeral arrangements, Lovelace demands of Belford, "to engage Tomkins to have every-thing ready for the opening and embalming. I shall bring Anderson with me" (8:46, letter 17, Mr. Lovelace to John Belford). Lovelace's quick retrieval of doctors prepared to perform an evisceration and embalming points to the obvious fact that eighteenth-century doctors had some amount of postmortem preservation knowledge. In fact, a large number of surgical books include chapters on embalming, and it can be assumed that most graduates of anatomy schools would acquire at least the basic methods of postmortem preservation, for both academic and mortuary purposes. Scientists and surgeons dramatically improved techniques, continually lectured on and published their findings, and contributed to the increased demand for embalming. Full arterial and cavity embalming is the seminal improvement in the eighteenth century; therefore, I would first like to summarize early intravenous techniques briefly. The great interest in anatomy and surgery during the Renaissance stimulated experiments with embalming methods. And because of the demand for medical men to have a more accurate knowledge of anatomy, "the problem of preserving bodies used for dissection was a matter of first importance" (Dobson 431). Both Leonardo da Vinci (1452–1519) and English physiologist William Harvey (1578–1657) experimented with arterial injections for the purpose of anatomical preservation (Dobson 431).[61] The first person to embalm by injecting a prepared preservative chemical solution into the blood vessels is believed to have been the seventeenth-century Dutch anatomist Frederik Ruysch (1638–1731), but his method, as that of many other embalmers, is unknown. The German scientist and physician Gabriel Clauderus, or more commonly Clauder (1633–1691), is believed to have used similar arterial-injection techniques to prevent cadavers from decomposing. British scientist Robert Boyle is credited with using "alkohol of Wine" to preserve a great green caterpillar in the late 1660s "as it were dry'd."[62] And there are numerous eighteenth-century treatises on how to delay or prevent putrefaction (the breaking down of flesh). One such example read to the Royal Society in 1750 is that of John Pringle, whose experiment includes mixtures of salt, camphor, alum, niter, and other ingredients used for modern embalming (Pringle experimented on beef shavings).[63] These experiments mark the progression of intravenous techniques that led to the Hunter brothers' groundbreaking discovery: full arterial embalming.

Eighteenth-century surgery books demonstrate that preservation techniques had become more widespread and publicized, and point to the persistence of the practice in the surgical field throughout the century, despite the infiltration of the undertaker. A brief overview of early English texts is helpful in assessing eighteenth-century innovations. The English translation of Philbert Guybert's *The Charitable Physitian* (1639)—which describes "the manner to Embalme a dead corps," including "to embalme the Heart"—marks the beginning of the innovation and publication of embalming methods (publication being rare and methods only vaguely described) (fig. 2.6). Guybert's method prescribes Venice turpentine and herbs with a series of balms to treat the corpse. Urns hold viscera following a specifically ordered dissection. (As we will see, Guybert's methods were later reformulated by surgeons such as William Hunter.) This treatise includes powder and liquid recipes, along with the estimate of six days for the heart to be embalmed, but it does not estimate the time for complete corpse embalming. One of the earliest English medical books to include a section on embalming methods is William Salmon's *Synopsis Medicinæ, or a Compendium*

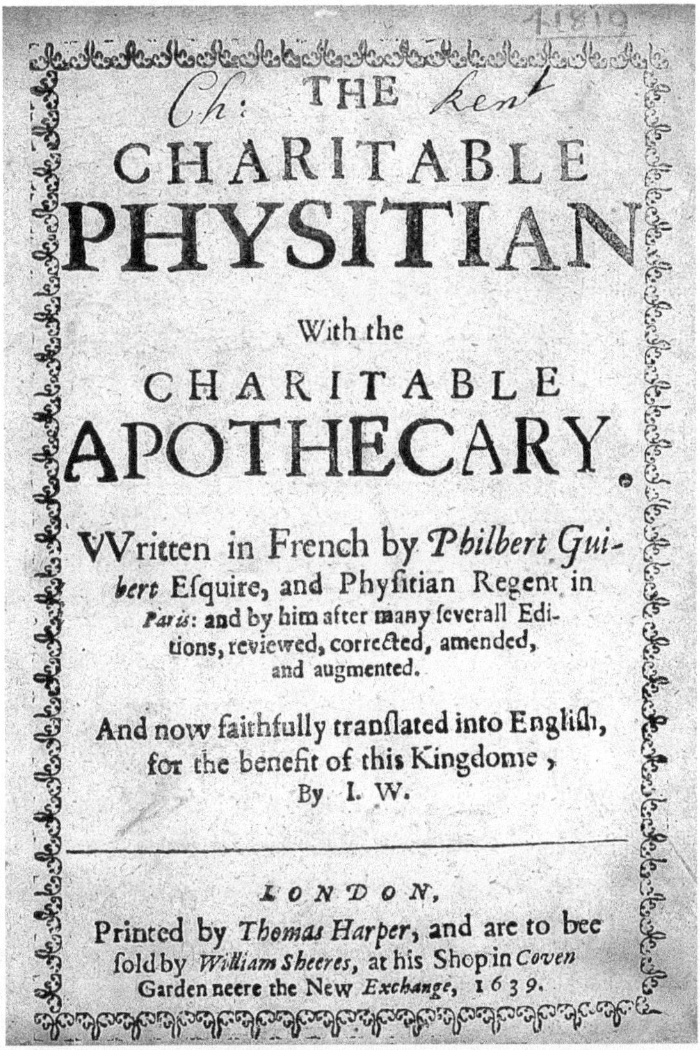

Ch: THE kent

CHARITABLE

PHYSITIAN

With the

CHARITABLE

APOTHECARY.

VVritten in French by *Philbert Guibert* Esquire, and Physitian Regent in *Paris*: and by him after many severall Editions, reviewed, corrected, amended, and augmented.

And now faithfully translated into English, for the benefit of this Kingdome, By I. W.

LONDON,

Printed by *Thomas Harper*, and are to bee sold by *William Sheeres*, at his Shop in *Coven* Garden neere the New *Exchange*, 1639.

Figure 2.6. Title page, English translation by I. W. of Philbert Guybert's *The charitable physitian with the Charitable apothecary*, printed by Thomas Harper, London, 1639. Public domain.

of Astrological, Galenical, and Chymical Physick (1671), which saw numerous editions through the end of the seventeenth century. In the 1681 second edition, Chapter XX, "Of Embalming," briefly explains the ancient Egyptian methods of embalming, then describes "The Modern Ways of Embalming," noting that the general method is "practiced through most of the European nations, and these are either without dissection, or with it."[64] Salmon notes that dissected bodies fare much better in the process (based on Egyptian methods), but he does include the process of "encearing," which is without dissection, and entails clysters (liquids injected into the lower intestines, similar to an enema) and chemical agents for cleansing the interior of the body: "wine vinegar, salt water, sea brine, or solution of niter, or a strong decoction of coloquintida" (393). Cerecloths (cloths soaked in wax or alum that covered the corpse's face) soaked in the same brine should then be wrapped around the body before it is placed in the coffin. An infamous dispenser, Salmon then advertises his business: "That the cerat for this purpose you may see in Our Doron or Supplement to Our London Dispensatory" (393). Salmon also provides the exact formulation for the cerat prescription, which includes beeswax, pitch, and frankincense. The section "The Modern way of Embalming with Aromaticks" includes dissection and bloodletting, methods later surgeons will also practice. Salmon directs that if the body is to be kept for a length of time before burial, incisions should be made into the larger vessels to better drain blood (this is an early form of intravenous embalming). Salmon also details the treatment of the body in spirits, its cleansing and drying, and then provides the prescription for a solution that the body is soaked in for eight to ten days before it is treated with the aromatic balm and wrapped in the cerecloth. "The Body being thus embalmed, will keep for some thousands of years without any decay," Salmon proclaims (396). Alexander Read's *Chirurgorum comes: or, The Whole Practice of Chirurgery* (1696) saw multiple editions throughout the late seventeenth and early eighteenth centuries. As with Salmon, Read first provides a summary of earlier methods and precise directions for his improved procedure (which is strikingly similar to Salmon's). Under what he considers "modern" embalming methods, Read describes the disemboweling, gashing, and bloodletting of the corpse, followed by the use of aromatic herbs, oils, and powders. The body is then anointed with turpentine that has been melted with chamomile oil and roses. He recommends that the body be wrapped in linen (ignoring the sheep's wool mandate of 1667), and "put into a leaden Coffin well soldered" (710). For the best result, he states that before burial, the body must be pricked with sharp bodkins so that "the Liquor preserving it from corruption may penetrate the deeper" (710). As with Salmon's, Read's method of pricking and soaking the body with alcohol predates intravenous techniques that later revolutionize embalming methods.

The first English text devoted to embalming is William Wilkins's *A Specimen of Embalming the Dead Intire without Disbowelling* (1690).[65] It is, indeed, this odd pamphlet that inspires Richard Steele's scathing portrait of undertakers in *The Funeral, or Grief-a-la-Mode* (1701). Wilkins argues that embalming "was esteem'd in all Ages, by all Nations, as appears throughout the Scriptures, and by the Works of many Authors," and that the "Art" is "acceptable to all People . . . especially by Persons of the highest Rank." His pamphlet consists of a brief explanation of his secret embalming procedures, which he claims to have perfected "by Ways and Means unknown to all Men but my self." Unlike previous procedures, Wilkins's method entails embalming without cerecloths or emboweling (the main concern of wary customers, who feared opening and mutilation). With the intention of producing a book-length study and leaving his embalmed collection "to Posterity," Wilkins presented his proposal, "A Petition of Mr. William Wilkins concerning the Embalming of dead Bodies," to the Royal Society in May 1710. The following month, *Philosophical Transactions* listed the entry "specimen of proposals for publishing the art of embalming dead bodies by William Wilkins," but subsequently the book was never completed.

Perhaps the most significant and widely published early book on the sole subject of embalming is surgeon Thomas Greenhill's *NExPoxHΔEIA, or, The Art of Embalming; Wherein is Shewn the Right of Burial, the Funeral Ceremonies, and the several Ways of Preserving Dead Bodies in Most Nations of the World* (1705) (fig. 2.7).[66] Here, Greenhill discusses theological, medical, and political reasons for embalming and includes numerous illustrations of Egyptian sarcophagi. This lengthy work defends the craft of the embalmer and, according to Gittings, describes embalming as "one facet of an increasing preoccupation with the preservation of the individual after death" (Gittings 104). It also reflects the growing public interest in Egyptology. In fact, Greenhill's *Art of Embalming* shows the usual method of bandaging based on knowledge gained from contemporary public mummy "unrollings," in which surgeons and anatomists publicly dissected mummified bodies in order to unmask the ancient secrets of death and preservation, but also to help demystify the superstitions associated with the corpse. As a form of advertisement for embalming, the book represents the interests not only of those prominent surgeons who already practiced embalming, such as Sir Hans Sloane (to whom Greenhill dedicates the book) and Francis Douce, but also of those who supplied the proper chemicals to surgeons, as noted in the lengthy list of subscribers.

Greenhill espouses embalming as a mortuary practice and seeks potential consumers by inciting fears concerning the disturbance of the corpse (he recommends embalming as an alternative to earth burial, where a corpse could be attacked by "voracious animals"). He goes so far as to paraphrase Sir Thomas Browne's *Hydriotaphia: Urne-Buriall, or a Discourse of The Sepulchrall Urns*

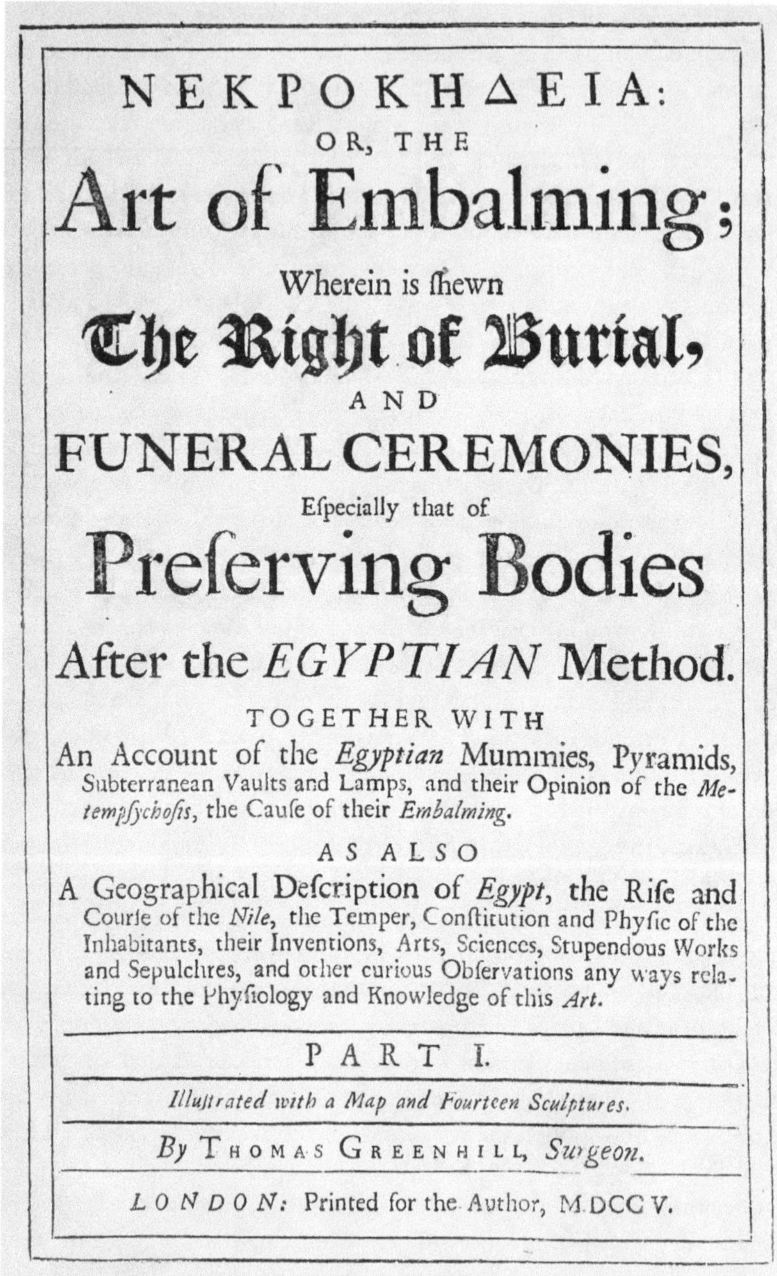

ΝΕΚΡΟΚΗΔΕΙΑ:
OR, THE
Art of Embalming;
Wherein is ſhewn
The Right of Burial,
AND
FUNERAL CEREMONIES,
Eſpecially that of
Preſerving Bodies
After the *EGYPTIAN* Method.
TOGETHER WITH
An Account of the *Egyptian* Mummies, Pyramids, Subterranean Vaults and Lamps, and their Opinion of the *Metempſychoſis*, the Cauſe of their *Embalming*.
AS ALSO
A Geographical Deſcription of *Egypt*, the Riſe and Courſe of the *Nile*, the Temper, Conſtitution and Phyſic of the Inhabitants, their Inventions, Arts, Sciences, Stupendous Works and Sepulchres, and other curious Obſervations any ways relating to the Phyſiology and Knowledge of this *Art*.

PART I.

Illuſtrated with a Map and Fourteen Sculptures.

By THOMAS GREENHILL, *Surgeon.*

LONDON: Printed for the Author, M.DCC V.

Figure 2.7. Title page. Thomas Greenhill's *Nekpokhdeia: or, The Art of Embalming* (London, 1705). Public domain.

Lately Found in Norfolk (1658), by stating, "Who knows the Fate of *his* Bones, or how often he is to be bury'd? Who has the Oracle of his Ashes, or where they are to be scatter'd?" (107). In fact, Greenhill's volume marks a point in history where social and medical attitudes toward the body and death were changing, intimating that it was natural to seek to preserve the bodies of the highly revered from corruption. Greenhill emphasizes this by remarking that God wants man to be preserved and uncorrupted for the final judgment: "Besides, we *Christians* ought to esteem *Embalming* a pious Work, acceptable to God, because it frees us from that Corruption which he so much detests" (113). What better marketing tool than to quote Psalm 16:10 and suggest that Christ was "*embalm'd* in order to his Resurrection" and was "exempt from the Laws of Corruption"; we should, therefore, follow his example.

Greenhill also defends the art of embalming from being practiced by charlatans and quacks, adding that "the noble *Art of Embalming* has been intirely ruin'd by the Undertaker."[67] Indeed, the occupational role of the undertaker had been emerging since the late sixteenth century. This rise in funeral contracting is evidenced in the annals of the barber-surgeons who for centuries exercised strict domain over the right to embalm and who in the seventeenth century took measures to see that their members were fit to carry out the practice of surgery, "of which embalming was an important subordinate skill."[68] In order to maintain strict control over embalming practice and to pass down methods and traditions, the barber-surgeons of London decided to apply for a new charter in 1604, the sixteenth clause declaring the "openinge searinge and imbalmeinge of the dead corpes to be pply [solely] belongeinge to the science of Barbery and Surgery, And the same intruded into by Butchers Taylors Smythes Chaundlors and others of mecanicall trades unskillfull in Barbery or Surgery, And unseemely and unchristian lyke defaceinge disfiguringe and dismembringe the dead Corpes."[69] These charter rights demonstrate the growing concern over the rise of the embalming trade by untrained wax chandlers, butchers, and tailors, possibly stemming from the expensive procedures performed by the barber-surgeons.[70] Yet by the end of the seventeenth century, the duties of the undertaker grew more complex, and they eventually assimilated the task of embalming, if in a crude form. In Richard Steele's scathing comedy *The Funeral, or Grief-a-la-Mode* (1701), an undertaker asks his assistants if they have "brought the Saw-dust and Tarr for Embalming," demonstrating the integration of the art into his duties (sawdust, tar, bran, and rosemary were inexpensive materials for lining coffins). In Richard Brinsley Sheridan's comedy *The Rivals* (1775), Sir Lucius O'Trigger asks Bob Acres if after the duel "would you choose to be pickled and sent home?" In the context of dueling and death, Sheridan also includes the malapropism Derbyshire "putrefactions " for "petrifactions," humorously signifying concerns about the state of the dead body.[71] And as the profession rose in importance, undertakers often displaced surgeons in embalming practice. In fact, in 1751, Frederick, Prince of Wales, was the first member of the British royal family to be embalmed by a professional undertaker.[72] Greenhill's *Art of Embalming* thus indicates the rise of the undertaker, the growth in the market for embalming (even among nonaristocrats), an increased concern over the dissolution of the body, and a literal interpretation of Christian-era preservation and bodily resurrection.

It is not until 1776 that a published account of improved embalming procedures appears. The Scottish anatomist William Hunter (1718–1783) is best known for his studies of the gravid uterus and is considered the founding father of scientific surgery (fig. 2.8). (Joshua Reynolds's portrait includes an example of the wet-tissue preparation of a uterus and fetus.[73]) It is therefore troubling to find that the innovation of embalming methods by William and his brother John is often purposely overlooked by recent biographers and medical historians, yet consistently present in funeral directing textbooks and histories of embalming.[74] The Hunter brothers, along with Charles White, their nephew Matthew Baillie, and others, lectured on and published their modernized versions of previously primitive methods of embalming. In fact, William Hunter is credited with being the first to report fully on arterial and cavity embalming as a way to preserve bodies for burial, and he was one of the few to make his method public. "Because Hunter was the first to report fully and openly the fluid and the method used, he is generally considered to be the originator of the injection technique of preserving human remains," state Habenstein and Lamers in today's standard mortuary textbook, *The History of American Funeral Directing* (99). After witnessing the unrolling of an Egyptian mummy, Hunter writes in his lecture "The Art of Embalming Dead Bodies" (1776), "Since that time I have often thought it would be pleasing if we could fall upon a method of preserving dead Bodies & I thought that mankind in general would wish to have the Bodies of their Friends &c Preserved. I conceive many People would be [pleased? ms illegible] with an Art of this kind or at least it would be useful to those who die abroad and are brought back home."[75] Hunter's discovery of complete arterial embalming, practiced at University College, London, and his private anatomy school at Great Windmill Street, involved injecting the body via the femoral artery with oil of turpentine, oil of lavender, oil of rosemary, and vermilion. He permitted the chemicals to diffuse throughout the tissues for several hours, then removed all the internal organs.[76] In fact, there are remarkable parallels between the techniques Hunter

Figure 2.8. Sir Joshua Reynolds, *William Hunter*, c. 1787. Oil on canvas. 125 × 100 cm. © Museums at the Royal College of Surgeons of England.

developed and those of the ancient Egyptian embalmers. Both eviscerated, and in the late period of Egyptian history, the organs were returned to the abdominal cavity, just as in Hunter's method. While the Egyptians placed the body on a bed of salt for dehydration, Hunter merely substituted plaster of paris.[77] Yet for Hunter, embalming meant a technique ensuring perpetual preservation of the corpse. His method involved permanent chemical arrest of decomposition while allowing for a premortem appearance.

William's brother John Hunter (1728–1793), the distinguished anatomist and surgeon, became known for his treatises on dentistry and venereal disease. John attended his brother William's anatomy classes and worked with him to improve embalming methods. Dobson notes that "John Hunter himself must have had no little experience in the technique, for he was engaged to embalm the body of the Earl of Moira in 1793" (438). The operation, which included a variation of William's process, was carried out by Everard Hone and John Doratt, one of Hunter's pupils, under Hunter's direct supervision.[78] In 1776, Hunter was honored by appointment as surgeon extraordinary to George III, who ironically refused to have his body "undergo any surgical operation . . . his majesty having in his lifetime always expressed an aversion to . . . embalming" (as recorded in the *London Morning Post* in 1821).[79] Notable as well is that depicted in the background of Joshua Reynolds's portrait of Hunter are the skeletal feet of Charles Byrne ("The Irish Giant") who also requested not to be dissected or preserved by anatomists, though in Byrne's case, his fears were obviously realized (fig. 2.9).[80]

Subsequent manuals and textbooks include the customary, primitive procedure of embalming through incision, perhaps suggesting that these cheaper forms of preservation were still desired. These are too numerous to be completely discussed here, but a few will demonstrate the continued interest in embalming procedures. Benjamin Bell's multivolume *A System of Surgery* (1783–1788) was published throughout the late eighteenth century, seeing multiple editions. Volume 6 includes brief chapters titled "Of Opening Dead Bodies" and "Of Embalming." Bell provides what he calls the "present method" of embalming, but shows he has not yet adopted the Hunter method, though he was a student of John Hunter. Instead of arterial embalming, Bell removes the organs, which are cleaned and then placed back in the body. Powders and oils fill the mouth and nostrils, in addition to the incisions, which are then sewn shut. If the body is to be taken a considerable distance, Bell recommends rolling all the extremities with bandages, and then covering the body with varnish (resulting in a mummified effect). He directs that two different colored cerecloths be applied, and the face of the corpse be painted by a professional, if preferred. This method reflects the Medieval and Renaissance practice, which emphasized the artistic presentation of the corpse.

Robert White's *Practical Surgery* (1796) is similar to Benjamin Bell's *A System of Surgery* (1788) in that it contains chapters titled "Opening a Dead Body" and "Embalming," and calls for preservative balms and powders. White, like Bell before him, states that the "surgeon is seldom called upon to perform this office, except upon the death of some great personage" (361). Unlike other surgery books, White's includes a recipe for the embalming powder he prescribes to fill the incisions:

Figure 2.9. Sir Joshua Reynolds, *John Hunter*, c. 1786. Oil on canvas. 140 × 111 cm. The skeletal feet of Charles Byrne, the "Irish Giant," are in the background. © Museums at the Royal College of Surgeons of England.

Powder for embalming

Lavender and rosemary flowers, each four pounds; the tops of wormwood, Arabian staechas, southernwood, with the leaves of the Syrian mastiche, aloes-wood, and calamus aromaticus, each three pounds; of the gums, myrrh, storax, Benjamin, frankincense, and the bark of sassafras, each one pound; nutmegs, mace, cloves, and cinnamon, each two ounces. Mix, and make into gross powder. (363)

The body's cavities are then "stitched up with the glover's suture" (361). He summarizes the work of Benjamin Gooch, whose methods were published in *The Chirurgical Works of Benjamin Gooch, Surgeon* (1792), which also directs that an opened corpse be properly sealed and clothed: "Mr. Gooch prepared his cerecloth with wax, resin, storax, and painters drying oil. The cerecloth can then be stained with vermillion. Varnish is used to seal the cerecloth and then a sheet is sewed on very close and smooth, with the finest herring-bone seam, then dressed and placed in the coffin" (362). It appears that later surgeons, such as Bell, White, and Gooch, emphasized the aesthetic properties of the preserved corpse, demonstrated by the attention to paints, seams, and colored cerecloths.

The publication of Matthew Baillie's lecture marks the last significant improvement in embalming methods. Baillie (1761–1823), a nephew of William and John Hunter (and brother of the famed author Joanna Baillie), was educated by his uncles and became famous as a physician and medical writer, best known for *The Morbid Anatomy of Some of the Most Important Parts of the Human Body* (1793). His lecture, titled "On the Embalming of Dead Bodies," given in Great Windmill Street on November 6, 1804, claims to make embalming "more easy of execution than any that, as far as I know, had been practiced by modern anatomists."[81] His uncle William's method, he wrote, "is certainly effectual in preserving a human body from decay for any length of time; but it is extremely tedious, occupying many hours, and is attended with much unnecessary difficulty."[82] Baillie modified the Hunterian method of embalming so as to provide the same results in a shorter period. In fact, his new process took approximately two weeks versus his uncles' methods, which took an estimated four to six weeks. After a brief history of Egyptian embalming and a summary of the Hunter method, Baillie describes his new, less tedious process. Using a solution of oil of turpentine, Venice turpentine, oil of chamomile, and oil of lavender, to which was added vermilion dye, he injected the femoral artery. He then opened the body and filled it with camphored spirits of wine. Like the Hunters, he rubbed the body with oil of rosemary or lavender and placed it on a plaster of paris bed. He embalmed the bodies of three young children by this simpler method and recorded that "all of them are still in the most perfect preservation, although they were embalmed more than twelve years ago; and from their present appearance I conclude, that they may continue free from decay for any length of time" (17). After Baillie's 1804 lecture, a British text dedicated to embalming does not appear again until the early nineteenth century, with no significant improvements recorded until new chemical discoveries were made (for instance, formaldehyde in 1867). Indeed, this eighteenth-century intravenous method is the early form of embalming procedure still utilized by surgeons and the mortuary business today. Together, this persistent publication and experimentation demonstrate both the demand for preservation and the rise of the "cult" of sensibility so vividly depicted in *Clarissa*.

Preserved and Circulated Curiosities

Investigating embalming in the eighteenth century, and how it fostered sentimental attitudes toward the corpse, would be incomplete without mentioning mummified "curiosities" and the practice of possessing, gazing at, and preserving the dead that *Clarissa* reflects. In *Philosophical Transactions* (1733–1734) Charles Balguy pronounces, "It is very common to see dead Bodies which have been preserved by Art for many Ages."[83] The British Museum, Royal College of Surgeons Museums, Wellcome Institute, and Surgeons' Hall Museums house thousands of wet and dry nonhuman specimens, many pathological, collected by eighteenth-century anatomists and surgeons. Sloane, the Hunter brothers, Read, White, Baillie, and others also maintained private collections of human remains, many of which were used as specimen study for anatomy instruction.[84] Evidence of impassioned Enlightenment interest in the human body, these collections also included eccentricities and biological oddities, ancient mummies, and modern examples of preservation techniques. Corpses were bought, sold, sawed, and placed in jars; they were in private rooms and displays, curiosities shown to invited guests as forms of morbid entertainment (such as the skeletal remains of Charles Byrne, discussed in Chapter 5). One notorious example involves Samuel Johnson's body. In *Loving Dr. Johnson*, Helen Deutsch makes a fascinating connection between Johnson's autopsy, his preserved lung, and an illustration in Baillie's *Morbid Anatomy* of "Emphysema." Uncovering that Johnson had been autopsied, Deutsch learns that "his lung, along with a kidney and a slice of his scrotum, had all been preserved," cataloged in the collection of anatomist (and assistant to William Hunter) William Cruickshank.[85] This practice of curiosity collecting is directly linked to sentimental mortuary practice: as science sought to uncover the mysteries of death, preserved human remains circulated throughout the period, in physical, rhetorical, scientific, and artistic forms. The distance between the art of the anatomist and the undertaker surely experienced an amount of slippage as fashionable mourning and sentimentality were on the rise.

As we have seen, ancient Egyptian mourning and burial rituals intrigued eighteenth-century eccentrics, who embraced Egyptian sepulchral culture; science and medicine explored various forms of modern embalming practices; and, as the British Empire spread to the east, Egyptomania and its materiality migrated to Britain.[86] While Greenhill's *Art of Embalming* unwrapped the mysteries of Egyptian embalming, experimenters criticized the Egyptian embalming methods as they worked to develop more successful techniques. Such criticism did little to dampen public interest in mummy unrollings, however; often these proceedings were witnessed by both professional and nonprofessional audiences and published in both professional and nonprofessional periodicals. The *Philosophical Transactions of the Royal Society* of January 1764, for example, includes "An Account of a Mummy, inspected at London 1763."[87] In a letter to William Heberden, John Hadley records that he invited several surgeons (including William Hunter) and a reverend to participate in the unrolling to be performed in his home. Given to him by the Museum of the Royal Society, the mummy was examined in order to investigate the "manner in which this piece of antiquity had been put together; to compare it with the accounts given of these preparations by ancient authors; and to see, whether there were any traces left of the softer parts; and, if so, by what means they had been preserved."[88] A few decades later, on April 10, 1794, Dr. John Blumenbach read his "Observations on some Egyptian Mummies opened in London," which details the opening and examining of several Egyptian mummies attended by the president of the Royal Society and many of its members.[89] Of the five mummies unrollings in 1782, three were done in the homes of private collectors who offered their mummies for investigation, and two of the mummies were from the Sloanian collection at the British Museum.[90] Four of the mummies were of children, and one believed to be that of a baby was discovered to be the skeleton of an ibis. Blumenbach's account is one of many; together they attest to the significant market for mummies by private collectors and public museums, as well as the circulation of preserved remains for medical and entertainment purposes.[91]

Also of interest, which Dobson briefly discusses in her essay, is the embalmers' own acquisition of English mummies, a fact recent biographies of eminent surgeons exclude. Many of the practitioners' experiments led to century-long preservations of "specimens" and to private natural history collections. These curious cases, which harbor, contain, and display the corpse, make the dead body less threatening by transforming it into spectacle. Smart Lethieullier, a prominent authority on Egyptian embalming and an antiquarian with a large personal museum, was an associate of the Hunter brothers who distinguished himself at the Royal Society on July 12, 1734, by presenting to its museum "the bodies of a set of triplets."[92] In a letter to Cromwell Mortimer, Lethieullier explains that the babies had died shortly after being born in 1714, and that he would like to donate the preserved corpses, "if they are thought worthy a place among the Human Curiosities." In the *Post Boy* for July 3, 1701, William Wilkins placed an advertisement for the public to view his "art" so that they might be convinced of the quality of his embalming skills:

> William Willkins over against the Mewes at Chearing Cross, well known and approved for his Art of Embalming; having preserv'd the Body of a Gentlewoman sweet and entire Thirteen Years without Emboweling, now to be seen: And he hath taken up the corps of Sire Edward Deering and Colonel Wharton in Ireland, after Nine Months putrefaction in the Ground. The Body of Colonel Corthope at Namure, and a merchant at Brussells, which he reduced to Sweetness without Emboweling, and their whole Bodies were seen and known by their Friends when brought to England. No Man performeth the like.[93]

Steele invariably read the pamphlet and quotes this passage in his preface to *The Funeral*, remarking, "He must needs be strangely in Love with this Life, who is not touch'd with this Kind Invitation to be pickl'd."

One of the most interesting personal collections was that of Manchester surgeon Charles White. A friend of John Hunter, he also attended William Hunter's anatomy class in London in 1748. As did John, Charles White collected "curiosities," the most notorious being the preserved body of Miss Hannah Beswick of Cheetham Hall—the "Manchester Mummy." Beswick was the daughter of John Beswick of Failworth, a man of considerable wealth, and White was her medical advisor for many years. White turned to Hannah to finance the hospital, and she lent him £30,000 in 1756. She died two years later, and White embalmed her (probably by arterial methods the Hunter brothers had recommended) and kept her body, preserved with tar, in a clock case in his house. (White also held the remains of Thomas Higgins, a notorious highwayman from Knutsford.)[94] Interestingly, Thomas de Quincey, who was born in Manchester in 1785, was taken to see White's museum when he was a child, and in his *Autobiography* he records that the mummy had been "placed in a common English clockcase, having the usual glass face; but a veil of white velvet obscured from all profane eyes the silent features behind."[95] After White's death in 1813, the mummy, which had been stored in an attic at his home in Sale Priory, was given, with the rest of his collection, to the Manchester Natural History Museum.[96]

John Sheldon (1752–1808) was responsible for a similar eccentric preservation; he had been apprenticed to a well-known anatomist and eventually took a post on the staff at a London hospital, where he met "Miss Johnson," who, supposedly, became his mistress."[97] It is said that she asked him to preserve her body, which he did in 1774, with the assistance of William Hunter, whose embalming lectures he had attended, afterward removing the embalmed and injected body to his own house.[98] As did John Hunter, he kept a "museum" of curiosities and showed visitors his mummified mistress. He supposedly told a friend, "She requested that I should make a mummy of her body, and keep her beside me—I have kept my word to her."[99] On Sheldon's death on October 8, 1808, Mrs. Sheldon donated the mummy to the Royal College of Surgeons Museum.

William Hunter's preservation methods also attracted attention after the 1775 embalming of Mrs. Martin van Butchell, an episode that Roy Porter and others have excluded from their scholarly discussions. Her husband, a London dentist and former anatomy student of William Hunter, heard of the Sheldon experiment and wanted to have his departed wife preserved. According to a memorandum written by Martin van Butchell, a copy of which is held at the Royal College of Surgeons Museum, he himself took part in the embalming proceedings, along with William Hunter, who details the procedure in "The Art of Embalming Dead Bodies," and William Cruikshank. Quite expensive and time-consuming, the embalming process lasted one month and cost more than 100 guineas (Kobler 139). They injected the body with preservatives to which coloring had been added to give Mrs. van Butchell's cheeks a rosy glow, and gave her a pair of "nicely matched glass eyes."[100] Van Butchell dressed and placed her in a glass-lidded case in his parlor and welcomed visitors. A vast amount of interest was, of course, taken in this "mummy," and, as Litten notes, the remarkable preservation inspired a ditty by Sir Richard Jebb, surgeon, often quoted by author Hester Thrale:

> To do his Wife's Corps peculiar honour
> Van Budgell [*sic*] wish'd to have it turned to stone;
> Hunter cast his Gorgon Looks upon her,
> And in a twinkling See [*sic*] the Thing is done.

William Hunter brought many visitors and other fellows of the Royal Society to view her at van Butchell's home. She was such a popular attraction that van Butchell eventually was forced to circulate a notice in the *St. James's Chronicle* (October 21, 1775), which reads,

> Van Butchell (not wishing to be unpleasantly circumstanced and wishing to convince some good minds that they [*sic*] have been misinformed) acquaints the Curious, no stranger can see his embalmed wife unless (by a Friend personally) introduced to himself, any day between Nine and One, Sundays excepted.

Van Butchell remarried, and subsequently the first Mrs. van Butchell was presented to John Hunter for his personal collection, and then donated to the Royal College of Surgeons Museum in London. Later, a writer described her body as a "repulsive-looking object," dry and shrunken. Mrs. van Butchell was joined in the Curio Room by the mummy of Sheldon's mistress in 1815. The mummies were exhibited together until destroyed by the aerial bombings in 1941.

In a more delicate instance, John Hunter was called in to perform the embalming of Princess Amelia, George III's elderly aunt. In his notes, Hunter records that the procedure took three hours.[101] He removed the abdominal organs and brain, which were placed in a lead urn, and the embalmed body was placed in a perfumed coffin. Completely bound in treated green cloth, the princess's body was wrapped in white and then purple silk, Hunter records that "The Purple is peculiar to the Royal family." Hunter biographer Wendy Moore remarks, "At last mummified and cocooned, the royal aunt was laid in her coffin and the lid soldered down, but not before certain privileged visitors had viewed Hunter's handiwork."[102] In a letter to Lady Ossory, Horace Walpole wrote, "She is already embalmed, cered, and coffined, her body is wrapped in I do not know how many yards of crimson silk, and she, they tell me, looks like a silkworm in its outward case."[103] Writing to his cousin Henry Seymour Conway (Anne Damer's father) from Amiens, September 11, 1765, Walpole describes another case of embalming that involved the Duchess of Douglas in Amiens: "You will not guess what she carries with her—Oh! nothing that will hurt our manufactures; nor what George Grenville himself would seize. One of her servants died at Paris: she had him embalmed, and the body is tied before her chaise:—a droll way of being chief mourner!"[104] In aristocratic fashion Walpole (d. 1797), who was the 4th Earl of Orford, requested in a codicil to his will, "I desire that my body be opened on my death" and then interred in the Walpole family vault at St. Martin's Churchyard, Houghton.[105]

Most notorious of embalmed *memento mori* are the remains of execution victims. It was not uncommon for the bodies of decapitated persons to be returned to their families for burial and preservation. Charles Bradford notes the case of Sir Walter Raleigh (d. 1618), whose wife had his head embalmed, "and kept it by her in a leather bag so long as she lived" (221n2). One of the most sensational and disturbing curiosities (rivaling the curiosity almost two centuries later of Jeremy Bentham's own preserved head at University College, London) is Oliver Cromwell's corpse. John Cranch's *Narrative Relating to the Real Embalmed Head of Oliver Cromwell* (1799) is one of several eighteenth-century pamphlets that details his postmortem "life" (fig. 2.10).[106] Cromwell died at Whitehall Palace on September 3, 1658. At the direction of the council, "the corpse was embowelled and embalmed on the following day."[107] Following lying in state and a grand state funeral, where effigies were used, Cromwell was buried in Westminster Abbey on November 10. In December 1660, the House of Commons voted to exhume the embalmed bodies of Cromwell, his son-in-law, and John Bradshaw. On January 26, 1661, they were removed from the Abbey and desecrated. Four days later, they were dragged to Tyburn and hanged for a day. Samuel Sainthill, a merchant who witnessed the hanging, wrote, "Cromwell in a green seare cloth, very fresh embalmed." The bodies were taken down, then the heads severed and displayed on poles at Westminster Hall as a warning against treason. It is not known how Cromwell's head was removed or by whom. The Cromwell Museum's website acknowledges that the head is known to have been exhibited in a private museum in London for the next twenty years, but we do not know its whereabouts after 1684. There is some evidence that it was in another private London museum as early as 1710. Cromwell's head was then purchased several times in the eighteenth century. Samuel Russell, a stage manager, had the head in the early 1770s, and displayed it in Butcher's Row, charging admission. He eventually sold it to James Cox, who at one time owned a private museum of curiosities and who bought it for £118. According to biographer Antonia Fraser, "He in turn sold it to three speculators for £230 and they exhibited it at the time of the French Revolution."[108] Cranch was then commissioned to write the pamphlet and draw the head for the exhibition at Mead-Court, in Old Bond Street. Cromwell's head may have been exhibited in other commercial museums until its sale to Josiah Henry Wilkinson in 1814, whose family gave the head to Sidney Sussex College, Cambridge, where it was buried in an unmarked location on March 25, 1960, ending his posthumous circulation.[109] In the eighteenth century Cromwell's head also served as a form of necropolitics: a preserved reminder of cultural and sociopolitical trauma following the Civil War and Interregnum.

Cromwell's eighteenth-century postmortem fame, as well as the persistent acquisition, private ownership, and public display of other *memento mori*, demonstrates the cultural obsession with preserved remains. Clearly, the eighteenth-century body was often preserved, and it became a form of spectacle, an object

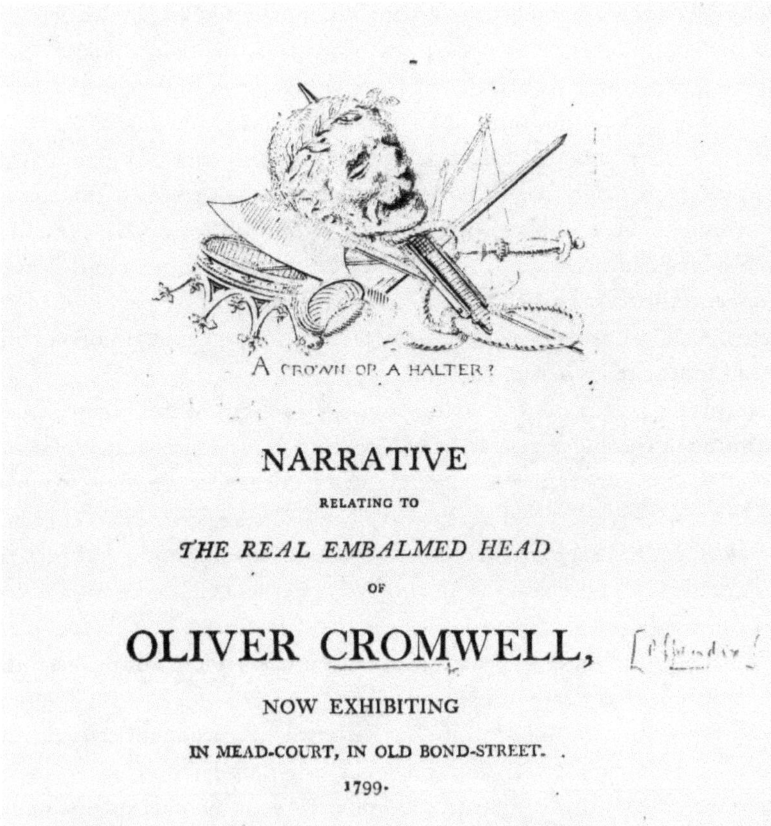
A CROWN OR A HALTER?

NARRATIVE

RELATING TO

THE REAL EMBALMED HEAD

OF

OLIVER CROMWELL,

NOW EXHIBITING

IN MEAD-COURT, IN OLD BOND-STREET.

1799.

Figure 2.10. Title page, John Cranch, *Narrative relating to the real embalmed head of Oliver Cromwell, now exhibiting in Mead-Court, in Old Bond-Street*, 1799. Courtesy of the British Library.

to be owned and observed by others. *Clarissa*, with its postmortem fetish, dramatizes this change in social hierarchies. It does not matter that, in the end, Clarissa's corpse is not embalmed, that Lovelace's necrophilic fantasies are summarily dashed. What matters is that we witness the *possibility* of dissection, evisceration, and preservation.[110] Lovelace wants to preserve Clarissa's beauty, bowels, and heart because he *can*. Her flesh functions as a relic, her preserved purity transformed into secular icon. Richardson includes this morbid scene in his novel of sensibility in part because the kind of embalming described was, indeed, practiced throughout the century. Most importantly, Lovelace's funereal orders for Clarissa's corpse embody a sociohistorical moment in which the fiction reflects fact: postmortem preservation was innovated, discussed, practiced, and circulated. As we have seen, it was a part of the larger shift in behavior that included using lead coffins, burying in vaults, preserving physical mementos of the departed, and embracing secular funeral arrangements. The question that this overlooked scene from *Clarissa* invokes, and with which this chapter commenced—"Were people embalmed in the eighteenth century?"—must be answered with a definitive "yes." Members of the aristocracy and wealthy middle and upper classes certainly had access to and sought out the preservation of their remains. And in Chapters 4 and 5, we will see how authors incorporated other forms of secular relics in their fiction.

Recovering the persistent discourse about, interest in, and performance of embalming in the eighteenth century helps to paint a larger picture of death's reception. Most importantly, I hope that this evidence has, in some small way, helped to shift perceptions of the history of embalming. Though never popular or frequent, embalming spread beyond royal tradition, infiltrating the burial rites of the aristocratic and wealthy in an effort to preserve social distinction posthumously. Embalming practice also demonstrates the increasing preoccupation with mortality and the corpse, and the intense anxiety about bodily dissolution and disruption after death. The scientific development of embalming not only allowed for better anatomical study, but directly influenced mortuary practice. Preserving, collecting, displaying, and circulating the dead body reflects a morbid sentimentality directly influenced by empirical thought and experiment. And as a new, empirical genre, the novel incorporates these practices and sensibilities in visceral ways. As *Clarissa* attests to, the novel is a living form, simultaneously creating and preserving the material. This fascinating and curious history of postmortem preservation techniques in eighteenth-century Britain ultimately contributes to the broader aim of *Death and the Body in the Eighteenth-Century Novel*: to rethink and reframe how we understand a culture's attitudes toward death, the body, and representation.

CHAPTER 3

Waxen Encounters

Flesh, Anatomy, and Autonomy

> Catherine had read too much not to be perfectly aware of the ease which a waxen figure might be introduced, and a supposititious funeral carried on.
>
> —Jane Austen, *Northanger Abbey*

Early in Richardson's *Clarissa* (1747), in a letter to Belford describing his account of the flight from Harlowe Place, Lovelace gives a peculiar description of Clarissa's face and skin:

> Her wax-like flesh (for, after all, flesh and blood I think she is!) by its delicacy and firmness, answers for the soundness of her health. Thou hast often heard me launch out in praise of her complexion. I never in my life beheld a skin so *illustriously* fair. The lily and the driven snow it is nonsense to talk of: her lawn and her laces one might, indeed, compare to those; but what a whited wall would a woman appear to be, who had a complexion which would justify such unnatural comparisons? But this lady is all alive, all glowing, all charming flesh and blood, yet so clear that every meandering vein is to be seen in all the lovely parts of her which custom permits to be visible. (3:27, letter 3)

This interplay between Clarissa as a live and a dead body, natural and unnatural, flesh and fleshless, points to her physical and mortal malleability. Doll-like and dead-like, Clarissa will later embody Lovelace's fetish during her comatose rape scene. Striking are the suggestions that Clarissa is skinless, her veins exposed in a visual dissection. This exposure borders the erotic: "every meandering vein is to be seen in all the lovely parts of her which custom permits to be visible."[1] While skin can be seen as a shield for an individuated, impermeable body, Lovelace's description penetrates beneath the surface, as if he has removed the skin. Also striking is the suggestion that her solidity confirms she is real: her "charming flesh" is "wax-like" in its "delicacy and firmness," which yields to the touch. Yet, her waxiness suspends Lovelace's reason: "flesh and blood I think she is!" While visual details and comparisons abound, references to touch (and resistance of the flesh) are evident and disturbing. This sensory, fleshy description not only embeds references to the state of her body during the rape, but foreshadows Lovelace's embalming fantasy (discussed in Chapter 2 of this book), where he desires to preserve and possess her flesh. Later, irritated at her refusals, Lovelace invokes a different wax metaphor and remarks, "But why will she break from me, when good resolutions are taking place? The red-hot iron she refuses to strike—O why will she suffer the yielding wax to harden?" (5:233, letter 23). His choice of words to describe her fear of intimacy not only includes sexual innuendo (BDSM practices of melting wax, of which Sade was a practitioner),[2] but again suggests the malleability of flesh: soft wax (or flesh) is yielding and pleasurable. As we have previously seen, Lovelace desires a parceling of Clarissa: hair, heart, bowels. These are "soft," morbidly "touchable" parts (if preserved correctly). Bronfen and others have shown that Clarissa's transparent "wax-like" flesh is uncannily similar to that of the Anatomical Venuses of the period.[3] "Every meandering vein" is exposed in these wax models, as in the Clemente Susini example from La Specola (fig. 3.1). Many wax models included painted wax veins, and some housed glass tubes that pumped red liquid through "the meandering veins." Later aligned with the undertaker in his request for her evisceration and heart, with this description Lovelace is aligned with the anatomist, closely dissecting, exposing, and preserving skin, flesh, and veins.

It has been well-documented that by the end of the century, wax had become a common component of the culture; the idea and form of wax circulated throughout the period.[4] Wax served as a malleable material for pseudopreservation; it allowed for anatomical models, cerecloths, and effigies. It also signified a bridge between the religious and medical worlds: wax *memento mori* and relics embedded in Catholic ritual emerged in Italian anatomy models; wax reliefs, seals, and death masks were a part of the growing mourning culture and undertaking trade. In their discussion of wax models, Hallam and Hockey observe that "Fusing anatomical realism with the symbols of *memento mori*, the object is am-

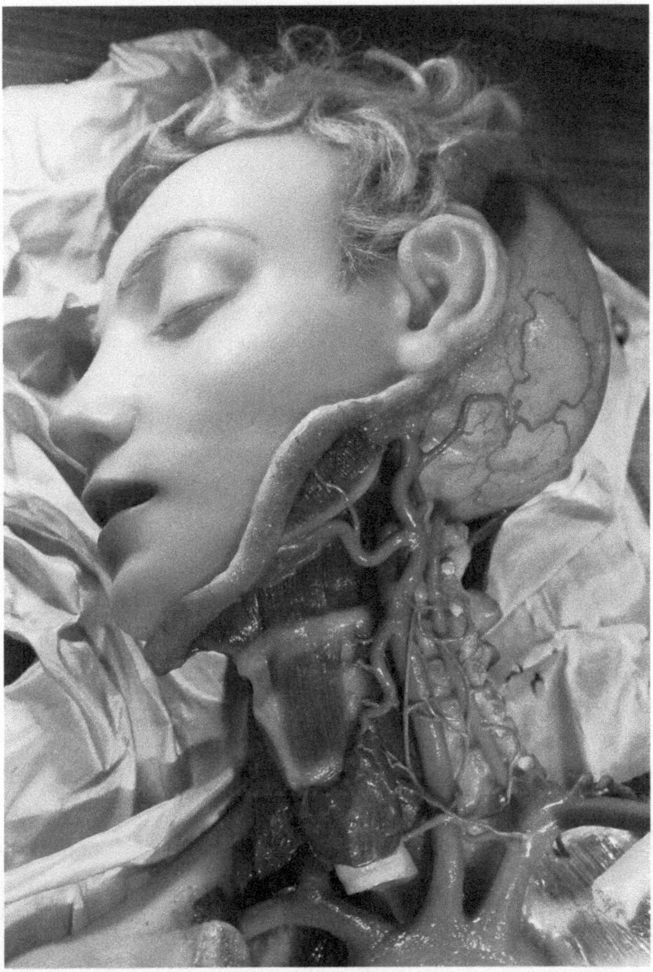

Figure 3.1. Wax anatomical model created by the workshop at La Specola, Florence (c. 1781–1786). Josephinum—Ethics, Collections and History of Medicine, MedUni Vienna, Photo: Ablogin.

bivalent in that it can be viewed as both an instrument of empirical instruction and a focus for spiritual contemplation" (65). The previous chapter's discussion of the preserved corpse helps to illuminate this chapter's examination of the role of wax in the history of *memento mori* (as dramatically figured in *Udolpho*) as well as eighteenth-century medical and cultural practices, such as the popularity of houses of wax and anatomical museums.[5] The closest element to flesh in the period, wax stands in for the dissected body and is central to the anatomist's art, thereby it plays an important role in the Enlightenment pursuit of knowledge. I suggest that in many ways this cultural exposure to specimens coincides with the undertaker's modes of preservation and the popularity of memorial tokens made up of human remains or signifying the body. It is no coincidence that the rise of the novel occurs alongside this public exhibition of human remains, dead bodies, and wax simulations. My final aim is to examine the role of wax and wax bodies in Ann Radcliffe's *The Mysteries of Udolpho* (1794). Specifically, I argue *Udolpho*'s heroine has a unique somatic response to the wax *memento mori* (that of a rotting corpse), signifying pleasure and autonomy (versus religious affirmation or moral example). With this, I suggest that not only does wax signify the empirical, tactile shift in the culture, but as it straddles religious *memento mori* and secular relic, the waxform loses its former Catholic association. I claim the role of wax in mourning and funereal practices, though ancient, is refashioned in the eighteenth century. As a flesh substitute, wax uniquely embodies the new sensibilities regarding mourning and the corpse.

Philosophical and Religious Wax

Wax has been a privileged figure in philosophical descriptions of the senses, thereby there is a lengthy tradition of invoking wax. The ancient philosophers described memory as "the permanent empress of a stylus on a wax tablet or of a signet ring on a wax seal."[6] In the seventeenth century, Descartes takes this up in his *Meditations* (1641), in which the metamorphosis of wax under a flame is the chosen metaphor for the distrust of sensory perception. In a well-known passage, Descartes describes the sensible properties of a wax ball and how they change as the ball is moved closer to heat.

> Let us take, for example, this piece of wax: it has been taken quite freshly from the hive, and it has not yet lost the sweetness of the honey it contained; it still retains somewhat of the odor of the flowers from which it was gathered; its color, figure, size, are apparent (to the sight); it is hard, cold, easily handled; and sounds when struck upon with the finger. In fine, all that contributes to make a body as distinctly known as possible, is found in the one before us. But, while I am speaking, let it be placed near the fire—what remained of the taste exhales, the smell evaporates, the color changes, its figure is destroyed, its size increases, it becomes liquid, it grows hot, it can hardly be handled, and, although

> struck upon, it emits no sound. Does the same wax still remain after this change? It must be admitted that it does remain; no one doubts it, or judges otherwise. What, then, was it I knew with so much distinctness in the piece of wax? Assuredly, it could be nothing of all that I observed by means of the senses, since all the things that fell under taste, smell, sight, touch, and hearing are changed, and yet the same wax remains.[7]

The malleable piece of wax allows Descartes to pronounce that "There certainly remains nothing, except something extended, flexible, and movable," and "no reasoning could aid in the perception of the wax or of any other body that would not—as all the same reasoning—better prove the nature of my mind."[8] Descartes sought to master materialism through reason and discipline; the melting, increased size, silence, and liquidation of wax signal the mistrust of the senses and all that is unstable in the physical world. Neil Hertz provocatively asks, "But why should such a drama be about a piece of wax? Descartes' commentators have no trouble explaining its nondramatic heuristic value to Descartes. Why wax? Because it can be easily described in terms of its impingements on all five senses, because it melts at a low temperature and thus lends itself to the experiment Descartes would perform on it, finally because it was ready to hand, an unspecial instance of an ordinary body, a common object."[9] Hertz then explains, "chunks of wax . . . were used for sealing letters, for receiving the imprint of the sender's signet ring—the use of wax that has made it, since the *Theaetetus* and the *De anima*, the privileged figure for the receptivity of the senses, or of the memory, or of the imagination or, more vaguely, of the mind" (175). Hallam and Hockey note the significance of the wax metaphor in Western traditions of describing the memory process: "The metaphor of memory as a wax tablet established connections between the body, sensory experience and material objects. The inscription of sensory perception upon the body and hence memory was likened to the imprint of wax seals achieved through the applications of signet rings."[10]

We see this metaphor arise again in Locke's thinking. In *An Essay Concerning Human Understanding* (1689), Locke uses the expression "white paper" to describe the tabula rasa theory that at birth the mind is a "blank slate." The early source for this theory is Aristotle, who in *De Anima* describes a "writing-tablet" on which nothing is written until it is traced by the mind. This Roman tabula—made of wax—was actually erased by heating the wax and then smoothing it over. According to Plato, perceptions and sensations are not ephemeral "but rather corporeal movements"; they assume and then "stamp a palpable shape on the susceptible wax."[11] Susan Stabile notes, "The point of a stylus or the carved signia on a ring transferred physical sensations (via the extremity of the hands) into mental impressions" (95). In his *Essay*, Locke's conception of the "blank slate" is thereby aligned with wax. In another well-known passage, Locke describes the causes of obscurity with the language of the wax and seal:

> The *cause of Obscurity* in simple *Ideas,* seems to be either dull Organs; or very slight and transient Impressions made by the Objects; or else a weakness in the Memory, not able to retain them as received . . . If the Organs, of Faculties of Perception, like Wax over-hardened with Cold, will not receive the Impression of the Seal, from the usual impulse wont to imprint it; or, like Wax of a temper too soft, will not hold it well, when well imprinted; or else supposing the wax of a temper fit, but the Seal not applied with sufficient force, to make a clear Impression: In any of these cases, the print left by the Seal, will be obscure.[12]

Sterne parodies this (as well as other numerous passages from Locke's *Essay*) with a bit of a vulgar reference in *Tristram Shandy* (1759):

> When *Dolly* has indited her epistle to *Robin*, and has thrust her arm into the bottom of her pocket hanging by her right side;—that that opportunity to recollect that the organs and faculties of perception, can, by nothing in this world, be so aptly typified and explained as by that one thing which *Dolly*'s hand is in search of.—Your organs are not so dull that I should inform you,—'tis an inch, Sir, of red seal-wax.
>
> When this is melted and dropp'd upon the letter,—if *Dolly* fumbles too long for her thimble, till the wax is over harden'd, it will not receive the mark of her thimble from the usual impulse which was wont to imprint it. Very well: If *Dolly*'s wax, for want of better, is bees-wax, or of a temper too soft,—tho' it may receive,—it will not hold the impression, how hard soever *Dolly* thrusts against it; and last of all, supposing the wax good, and eke the thimble, but applied thereto in careless haste, as her Mistress rings the bell;—in any one of these three cases, the print, left by the thimble, will be as unlike the prototype as a brass-jack.[13]

Locke's chosen metaphor for obscurity is taken up by Sterne for parodic purposes, but it also plainly deciphers his meaning while signaling the ubiquitous materialism of wax in eighteenth-century literature and culture. As Enlightenment debates about the separation of the mind and body flourished, wax seems to be an appropriate metaphor and medium for the corporeal.

Along with its presence in philosophical discourse, wax has been a material staple of religious rites and relics. In England, for example, the Worshipful Company of Wax Chandlers was established in the fourteenth century.[14] Beeswax candles produced by chandlers were a necessary element for funerals, and churches were decorated with wax reliefs and figures. Used for votive and *ex voto* images and figures, wax offerings were ubiquitous in pre-Christian and Christian times. Sean Silver notes that in addition to ampullae of saints' blood purchased by pious pilgrims to Canterbury Cathedral, other votives were curiosities such as "fish bones, bodily excrescences, crutches, shackles, and thousands of wax images of diseased limbs and organs" (172). After the Reformation, "Protestant objections to polychrome sculpture were emblematic of a much larger rejection of Catholic practice, a doctrinaire distaste that lingers in Western thinking on art today."[15] During the English Civil War, "Puritans smashed wax idols and drove wax modelers in England underground."[16] Polychrome was a Catholic practice, and all the more so if the figures were in wax. In 1672, the diarist John Evelyn, writing in tense times, connected the display of wax figures to what he viewed as attempts to restore Catholicism to Britain: "I went to see the fopperies of the Papists at Somerset house and York house, where now the French Ambassador had caused to be represented our Blessed Saviour at the Pascal Supper, with his Disciples, in figures and puppets made as big as life, of wax work, curiously clad, and sitting round a large table, the roome nobly hung and shining with innumerable Lamps and Candles, this exposed, to all the world, all the Citty came to see; such liberty had the Roman Catholics at this time obtained."[17] This French display was seemingly an act of religious provocation, designed to have a broad impact. Luke Syson notes that "Many in that period, as well as before and since then, believed that wax, a natural material with transformative potential, was endowed with supernatural properties, and the use of wax was often connected to the fashioning of witches' poppets" (26). Syson also reminds us that in John Webster's play *Duchess of Malfi* (1613), the duchess is forced to look at wax figures of her husband and children, portrayed as if dead. Later, we will see these anxieties and provocations dramatically figured in the waxwork in *Udolpho*.

As Evelyn points to, in the Early Modern period, with superstition or belief still effective weapons of propaganda, waxworks were created by the political elite for popular use and circulation. Until the execution in 1649 of Charles I made it impossible to continue the practice, the full-length, crowned, and gowned effigies of the English kings and queens were displayed recumbent on the lids of their coffins so as to encourage the crowds to pray for their immortal souls.[18] By the end of the sixteenth century, the figures were made from wax (most likely based on Italian models) (Syson 26). Throughout the eighteenth century, funereal effigies remained a part of the burial practice for the aristocracy and signaled "the move from the religious to the secular in mourning and death culture."[19] Matthew Craske describes a notable example: "Instead of commissioning a marble monument to her dead sons, the Duchess [of Buckingham] decided to have them modeled in wax. Shortly after her own death in 1743, a series of wax figures were set up before the monument: those of the Duchess, accompanied by Robert Marquis of Normanby (who had died in 1714), were placed beside a recumbent effigy of Edmund" in Buckingham House.[20] Later this effigy became part of the display in Westminster Abbey, illustrated in its catalog, *A View of the wax work figures in King Henry the 7th's Chapel, Westminster Abbey* (1769), which includes illustrations for several of "the great personages whose effigies are here represented." Prices for waxworks remained consistently high over time. In 1702, for example, Frances Stuart, Duchess of Richmond and the former mistress of Charles II, paid Mrs. Goldsmith an enormous £260 to fashion her funeral figure (and even had her own parrot stuffed to go with it). The sculpture took nearly a year to make. In another strange case, rumors circulated that Lady Godolphin, daughter of the Duchess of Marlborough, found the death of William Congreve quite intolerable and supposedly placed "a wax figure of the writer at the table every day as an invited guest and to take the same to her bedroom at night."[21] Though eccentric, these wax examples signal the cultural shift from the religious to the secular, from the spiritual to the material.

Entertaining Waxworks

Dolls (generally imported from France), figurines, effigies of royalty, and statues of saints were still modeled in wax in the eighteenth century. As Lady Godolphin's waxwork of Congreve symbolizes, wax was entering the private and public spheres in unique forms. In a quotidian example, we know that by midcentury the upper class were replacing tallow candles (which have a bad odor and do not burn clean) with more expensive wax candles.[22] We see an even earlier seventeenth-century example in Samuel Pepys's diary. In his December 15, 1664, entry, Pepys remarks, "This night I begun to burn wax candles in my closett at the office, to try the charge, and to see whether the smoke offends like that of tallow candles."[23] In Burney's *Cecilia* (1782), the nouveau riche Harrels prepare for a party by placing "wax candles in every room" (at what would be quite an expense).[24] Beeswax is so ubiquitous that it enters women's conduct books early in the century. For example, in *The lady's delight, Or Accomplish'd Female Instructor: being a very useful companion for ladies, gentlewomen, and others . . .* (1740), the author instructs women to make waxworks, including candles and figurines, and preserve various items with wax, such as flowers.[25] One section describes how to make life masks and "To make or frame curious Figures in Wax-work" (161–64). In Charlotte Smith's *Ethelinde, or The Recluse of the Lake* (1789), the eponymous protagonist, on the verge of destitution and suffering from the loss of her father, is compared with a waxwork by the ridiculous Miss Newenden: "Lord! child, what's the matter with you? you look like a wax figure, with your fixed heavy eyes and pale face. Come prythee have a little more spirit . . . why you are fit for nothing."[26] Here Smith signals that her middling and upper-class characters as well as her readers have visited wax houses or seen wax dolls and effigies, thereby fully understanding the comparison.

As we saw in Chapter 2 with Bentham's auto-icon and Cromwell's embalmed head, various forms of wax bodies circulated in eighteenth-century popular culture. Julie Park recognizes in *The Self and It: Novel Objects in Eighteenth-Century England*, "Other spectacles of humans transformed into commodified and consumable objects arose in the popularity of dolls, waxworks, and automata as forms of entertainment, thus demonstrating the period's fascination with 'man-made' versions of the human, as well as objects made to look like the human."[27] In their uncanny replication of the human form, and often a *particular* human form, waxworks symbolize the "growing complexity of modern subjectivity" (Park xv).

In a well-known example of the confrontation with human-like, mechanical forms from Burney's *Evelina* (1778), the eponymous heroine, along with relatives and friends, visits Cox's Museum in London, jeweler James Cox's actual collection of mechanical toys, automata, and various curiosities.[28] The frontispiece to the museum's catalog includes a profile of the head of George III, eerily resembling Curtius's wax heads of guillotine victims produced later in the century (fig. 3.2). In addition to a pineapple that suddenly opens to reveal a nest of singing birds, as well as a "concert of mechanical music" that emits from an unknown source, Evelina and company would have seen exotic and intricate automata. The 1773 catalog lists the "Automaton in the habit of a Chinese," "two Automaton figures of a man and woman, in Turkish habits," and an "Eastern Lady" that moves "from right to left" while holding a guitar.[29] Overwhelmed by the exhibition, Evelina describes it as "very astonishing, and very superb; yet it afforded me but little pleasure, for it is a mere show, though a wonderful one."[30] Others describe

Figure 3.2. Print, frontispiece to the descriptive inventory of Mr. Cox's Museum. Depicts wax head of George III (c. 1774). © The Trustees of the British Museum.

it as a "brilliant spectacle," yet Evelina is puzzled, admitting "I seem to miss something" (76). Perhaps the emptiness of the automata intrigues and perplexes, confounding her reaction to mechanization and commodification. Park observes that similar to the effect of dolls, waxworks can "complicate notions of agency and mastery in fashioning the self" (xv) (as we see in Evelina's reaction to automata), but their prevalence can also activate this agency and mastery, instigating a construction versus objectification of the self.

Waxworks and automata circulate as public forms of entertainment and intrigue as early as the late seventeenth century. The 1664 advertisement for the

London fair exhibition *Extraordinary, and unparallel'd curiosities, composed of wax, (so near the life, that they cannot be excelled by art) in seven several scenes* describes biblical figures and scenes ("Virgin Mary," "Angel Gabriel") as well as mythological ones: Neptune, Jupiter, Venus, and Cupid "riding upon a Dolphin."[31] The advertisement notes that "If any Gentlemen or others desire to have their Pictures in Wax, if they please to repair to these Artists, they shall have content." Evidently the artistic and entertaining, as well as religious and secular, modes of wax were circulating as early as the mid-seventeenth century. And as Laura Engel reminds us, "Even before Madame Tussaud arrived in England from revolutionary France in 1803 with suitcases full of wax bodies, 'Mrs. Salmon's Royal Waxworks, comprising some 140 life-size figures "all made" by her own hands,' had been attracting visitors since 1693."[32] The advertisement from 1763 boasts a collection of more than two hundred "Figures as large as Life" filling four rooms. These rooms contained numerous scenes of rulers, heroes, and heroines of British history and folklore, as well as prosaic social types. Park notes that figures in Salmon's collection included "Queen Elizabeth, with Lady Margaret Russel, who pricked her Finger and bled to Death in one room; King Henry VIII introducing to Court Anna Bullen, to the great dislike of Queen Katherine and Cardinal Wolsey; the chaste Nuns of Collingham, who slit their Noses and upper Lips to preserve their Virgin Vow; a Cherokee King, with his chief; Peter the wild boy; the British Giant; an Old Maid and her Sweetheart; a Dutch Christening; a fine Representation of the Death of Werter, attended by Charlotte and her Family; and Mrs. Salmon herself, with three of her Children."[33] Perhaps the most fantastic spectacle in Salmon's collection of waxworks was "a beautiful Rock, ornamented with Pearls, Corals, and rich Stones" that contained "six Caves, in which" one could see "Hermits moving, Mermaids waving." In fact, as picturesque gardens grew in popularity, wax hermits were placed in caves in extravagant estate gardens (wax hermits were much cheaper than hiring live actors).[34] Salmon's collection was so popular that it is referred to in *The History of Sir William Harrington* (1772), an epistolary novel by Anna Meades (and possibly revised by Samuel Richardson). When describing a grand home she is visiting, Miss Harrington writes to her friend detailing the portrait gallery: "You have seen the wax-work in Fleetstreet, Cordelia, and doubtless remember the beautiful figures of Mrs Salmon and her infant child? Here is one of the Countess's drawn in the same manner at full length, sitting in a chair, with a daughter of about three months old, the sweetest little soul, lying in her lap. Such pleasure! such tenderness! in the eyes and whole deportment of the mother. It is reckoned, and I really believe is, one of the best pieces, perhaps, in the whole world. Lord C.

and my sister, are to sit for their pictures when next in London. So much for the gallery."[35] Another popular exhibit in Paris catered to English tourists. The collector's label for *Partie du Cabinet d'Aubin* (c. 1780) uses both French and English, which suggests the gallery was visited by the English, to whom Aubin sold small wax models as souvenirs. The print lists some of the gallery's exhibits, such as "wax death masks, busts and hands, alongside a fashionable Chinese grotto."[36] Whether erected as "lively" representations of history, homage to recently departed members of the royal family and military heroes, or visual mediations of beloved literary characters, the waxworks of eighteenth-century Britain were valued for the way they reconstituted life as a purely material phenomenon (Park 96).

This materialism and individualism is noted by Madame Tussaud biographer Kate Berridge, who observes that "Her long life saw a general cultural shift interest away from posthumous glory in heaven to a greater preoccupation with self-definition on earth."[37] Waxworks (a three-dimensional form of portraiture) reflect a human form of "fleshy desire" (Engel 296) reflected in anatomical models, which rise in popularity simultaneously with celebrity houses of wax. Women have a lengthy history with wax, as Marjan Sterckx reminds us: "Women's close association with modeling, especially wax modeling, dates back at least to the Middle Ages, when nuns made candles, wax flowers and small statues of saints and the Virgin Mary for convents and private chapels. This tradition continued through the seventeenth century."[38] Before Mrs. Salmon, Patience Wright, an American in London, exhibited in her own "house of Fame" from 1772. Wright was known for her "uncanny living likenesses of celebrities, including George III."[39] Catherine Andras sculpted portraits of the aristocracy and royalty (including Lord Nelson) and was appointed in 1802 as "Modeller in Wax to Queen Charlotte." Apprenticed to the sculptor Philippe Curtius, stormer of the Bastille (and her guardian), Marie Gresholtz (later Tussaud) produced effigies of the royal family's decapitated heads, as well as the heads of many other victims of the Revolution, such as Marat.[40] Sleeping Beauty (c. 1767) was based on Louis XV's mistress Madame du Barry and is one of Curtius's earliest works (and the oldest specimen now existing in the Tussaud collection).[41] Sleeping Beauty can be seen as what Joanna Ebenstein terms "the product of a male imagination" and "veiled fetish."[42] In the late eighteenth century, mechanisms were even installed to feign breathing and chest movement. She is breathing yet unconscious; the physical product of male voyeurism. In *Phantasmagoria: Spirit Visions, Metaphors, and Media into the Twenty-first Century*, Marina Warner notes "The illusion of permanent sleep is invoked to deny the reality of death . . . The Sleeping Beauty functions as anti-memento mori . . . she promises immortality as the suspension of time."[43] As we will see, she opposes the effect of Zumbo's decomposing corpses and van Rymsdyk's dissected Venuses, which violently pronounce mortality, yet it is her waxy fleshiness that seduces. Warner explains, "Such a sculpture conveys a seductive vision of erotic, feminine catalepsy, which the peculiar translucence and slight sweatiness of the wax medium suit so creepily."[44] It is this fleshy medium that is of interest here: wax powerfully conveys the materiality of the corpse without putrefaction.

Anatomical Wax

Understanding the mysteries of the body involves the tearing down and preservation of flesh. John Hunter's vast collection of wet specimens prepared in alcohol and spirits and housed in the Hunterian Museum reminds us that gross anatomy and pathological examples were integral tools for the education of eighteenth- and nineteenth-century surgeons. Collecting, boiling, and preserving flesh was the morbid practice of the anatomist.

Whereas entertaining waxworks were circulating in eighteenth-century culture, waxworks were being produced in the seventeenth and eighteenth centuries to fill gaps in anatomy classes since the supply and storage of corpses was problematic. Seventeenth-century Dutch anatomist Frederik Ruysch, whose arterial experiments commenced modern embalming methods, kept a notorious cabinet of preserved preparations. Many of these preparations in the 1690s involved vascular injections with fluid wax (and later Ruysch included pig's blood as a red "dye").[45] Steven Jay Gould explains, "Ruysch made about a dozen tableaux, constructed of human fetal skeletons with backgrounds of other body parts, on allegorical themes of death and the transiency of life. . . . Ruysch built the 'geological' landscapes of these tableaux from gallstones and kidney stones, and 'botanical' backgrounds from injected and hardened major veins and arteries for 'trees,' and more ramified tissue of lungs and smaller vessels for 'bushes' and 'grass.'"[46] A pioneer in the art of preserving the human body, his morbid tableaux of skeletons and waxen body parts intersected science and art.

The influence of Ruysch and others who implemented wax injections to help preserve wet specimens is seen throughout the period. Bronfen has pointed out that in the eighteenth century, the use of wax models increased for anatomy students, providing access to the body without requiring contact with decaying flesh. She states, "Producing a substitute of the corrupt and putrefied dead body that would mask death, these models are endemic to a general cultural effort to eliminate the impure state of mutability and decay by replacing it with a pure and immutable wax body double" (99). The transactions of the Royal Society include numerous descriptions of wax specimens, as well as physical examples of injected wax specimens presented to the Royal Society. In April 1725, John Ranby presented the society with two preparations of a human fetus "which he had injected with red wax to highlight the skin's fine vessels."[47] A shortage of cadavers induced Abraham Chovet (demonstrator in anatomy to the barber-surgeons) on March 24, 1726, to offer the society his proposals for "making Anatomies of human bodies in colourd Wax . . . exhibiting the circulation of the blood by a red Liquor running thro' Veins and Arteries of Glass," as well as a wax anatomy by itself (Appleby 14). The anatomies were subsequently demonstrated at a June 1732 meeting and displayed in December 1733 at "Mr Lamark's, Surgeon, in Orange Street, Leicester Fields." Guerrini notes that one waxwork represented "a woman chained down upon a table, suppos'd opened alive." Five shillings were charged for a viewing, and female visitors would be attended by "a Gentlewoman qualified" ("Anatomists" 236). On February 1, 1798, John Abernethy read "Observations on the Foramina Thebesii of the Heart," which describes injecting "different coloured wax which had been impelled into the coronary arteries and veins," allowing for locating diseases in the cavities of the heart.[48] And William Hunter's wet mounting of specimens, still housed at the Hunterian Museum, relied on wax injections, among other techniques.[49]

These wax innovations fix the condition of the body, suspending it for study. Hallam and Hockey recognize this significant form of stasis: "Unlike the static marble bodies rendered as memorial effigies dedicated to identifiable persons, anatomical waxes were models of anonymous bodies used to provoke a different kind of remembering. Wax models were bodies without lived histories or identities in that that they retained no distinctive traces of the deceased persons from whose physical components they had been fashioned. Thus, it is not persons that were being 'fixed' in memory, rather it was a generalized knowledge of human anatomy that was being imparted in an educational context" (67–68). Whereas wax anatomical models are depersonalized auto-icons, preserved for detached study without the contamination of rotting flesh, later I will demonstrate the slippage of wax into religious, emotional, and sentimental territories.

By the early eighteenth century, European anatomists and modelers were creating entire wax bodies (with human skeletons as their basis) and body parts for teaching purposes. Until the eighteenth century, the poses of the figures in anatomical atlases aligned them with orators or mythological heroes; the mostly male bodies featured in atlases stand or kneel like macabre versions of Greco-Roman or Renaissance statuary.[50] For example, the full-body male wax model now in the Hunterian collection (fig. 3.3) is posed in an ancient Greek manner. (It is actually a wax cast of a flayed criminal that William Hunter used to show artists what lies beneath the surface of the skin.) Enlightenment wax models often had their

Figure 3.3. Michael Henry Spang, écorché figure, c. 1761. Wax, cast from flayed criminal. Bequest to William Hunter, 1783. © The Hunterian, University of Glasgow.

eyes open and appeared strangely alive. As numerous scholars, to include Elisabeth Bronfen, have recognized, by midcentury female models took on a supine and morbidly eroticized appearance, posed, exposed, and lifelike.[51] The male-dominated science of obstetrics was replacing the art of midwifery (a replacement that William Hunter was the center of), and intricate wax models depicting the various stages of gestation engendered this shift. Hunter's *Anatomia Uteri Humani Gravidi Tabulis Illustrata* (*The Anatomy of the Human Gravid Uterus Exhibited in Figures*) (1774) was based on the dissection of preserved, pregnant cadavers, and several artists, including Jan van Rymsdyk, created anatomically correct illustrations of Hunter's work for the book. Hunter had the artists draw during the actual dissection process, but he also utilized wax models to assist in the process, which was the case for most anatomists.[52]

While the erotics of female waxworks have received much attention, my broader interest is in the physical circulation of medical models.[53] Wax allows for ceroplastic productions: they can be stored, handled, and manipulated with little deterioration. For example, Susini's "Medici Venus" or "Demountable Venus" (1780–1782) at La Specola can be disassembled into seven anatomically correct layers (eventually revealing a fetus curled in her womb."[54] A "perfect embodiment of the Enlightenment values of her time," "the pinnacle of divine knowledge," the Medici Venus was highly regarded by anatomists and "copies were commissioned for a variety of museums and teaching collections around Europe" (Ebenstein 24). In 1734, "Paris-trained anatomist and modeler Abraham Chovet exhibited a female depicted as though in the painful process of being vivisected" (Ebenstein 37). His mechanical wax model displayed a pregnant woman cut open to reveal the undelivered baby, with red fluid pumping through glass tubes to simulate the circulatory system in pregnancy. (Later, Giullaume Desnoues's and Abraham Chovet's models would be bought by Rackstrow.) Bronfen, Sander Gilman, and other scholars previously recognized that these scientific specimens served to "cover distance and control both sexuality and death by rendering the mutable, dangerously fluid, destabilized feminine body in a cleansed, purified, immobile form" (Bronfen 99). Yet we must keep in mind that the male form was also displayed (albeit in less erotic poses, and without hair, makeup, and jewelry). I am not seeking to dismiss these influential Freudian and feminist readings of waxworks, but by resituating human remains and anatomical renderings we can better understand how the dead body becomes an object appropriate for circulation, parceling, and ownership. The Hunter brothers, Rackstrow, and numerous other collectors understood that the sensibility of the period was shifting as science exposed and preserved every possible human part. Since dissecting theaters and private collections were not accessible to all, public anatomy museums democratized the body (and allowed female visitors). In fact, nearly forty anatomy museums were created in England between 1739 and 1800, and with the popularity of lectures involving real cadavers and wax displays, palpable human remains were in demand.[55] Classifying bodies and diseases, and displaying the abnormal (morbid anatomy), depersonalized the process of understanding the human. Pathologists "dismembered the dead body and preserved the fragments, whether by injection or by storage in fluid, fashioning them into material culture."[56] Visual experience of dead bodies was integral to the teaching of anatomy; this involved drawings and illustrations, dissection theaters, wax casts, and wet specimens. While the classroom experience was visual, it was the experience of the hands, of touching, that coordinated with sight for a full understanding of the body.

Anatomical models (male and female alike) invoke a morbid pleasure received through the suspension of decay; the waxforms resist to touch, are tactile and mobile (limbs can be manipulated, forms posed, parts removed). Ebenstein recognizes that wax models represent the intersecting ways Catholicism and medicine "seek to preserve and effigize the body" (18), and since the major collections were housed in European, mainly Catholic countries, we can understand this link with iconography and anatomy. For example, Benedict XIV employed wax sculptor Ercole Lelli to sculpt a comprehensive anatomical display in wax which made Bologna the great center for the wax production of anatomical models.[57] Yet for the English, mainly Protestant, culture, these anatomical icons would be generally devoid of contemporary religious context. Beliefs founded on faith were being questioned in the Enlightenment, and empirical observation, as the wax anatomy models signify, was shifting Western perspectives.

This secular interpretation of wax models is witnessed in terms of their circulation in "profane" venues. Discussed in Chapter 2, most of the period's anatomists collected their own curiosity cabinets, many containing wet and dry human specimens, as well as wax models. Enlightenment private and public collections helped shape knowledge and developing notions of the body and the individual.[58] Many sourced from unknown criminals, dissected and preserved specimens often depersonalized the art of the anatomist, and turned the body into a morbid form of material culture. Richard Barnett recognizes, "Anatomical museums, lectures, and demonstrations became a fashionable diversion, part of a pleasure circuit which took in the Tower of London and the grand boulevards of the West End" (366). *A catalogue and particular description of the human anatomy in wax-work, and several other preparations* (1736) is an early example of a public anatomy collection exhibited at the Royal Exchange. Wax sculptor Benjamin Rackstrow's Museum of Anatomy and Curiosities was a public collection that displayed "anatomical figures, and real preparations, also Figures resembling Life."[59] One of the surviving catalogs from 1782 overviews some of the contents and gives insight into contemporary tastes. The Anatomical Collection lists specimens preserved in spirits, many of human remains ("monstrous births," "diseased wombs"), as well as waxworks, including molds taken from life of the recently deceased Mr. Bamford ("the Staffordshire Giant") and Mr. Coan ("the Norfolk Dwarf"), as well as wax models of the human reproductive system. These anatomical curiosities were juxtaposed in an adjoining room by waxworks of "His Late Majesty George the Second, in His Parliament Robes," life masks of the Prince of Wales and the Late Marquis of Granby, next to the embalmed body of "A real Ancient Mummy." In this same room were exhibited "Exquisite Wax-Models" of "A Tinker and His Family, &c. represented in Miniature," "A Young Lady with Children gathering Apples," and "An Elderly Lady with Children, a Cat and Kitten, &c. surprised and discommoded by the sudden Appearance of a Gentleman, and Grey-hound." Shocking for the audience was most likely the numerous wax figures of women with wombs dissected, one "exposing the Child and Placenta" with a "Navel-String passing around the Child's Neck." His *pièce de resistance* was Chovet's "wax sculpture of a pregnant woman, partially dissected, with claret running through glass tubes representing the circulation of her blood."[60] And as Craske and other historians have illuminated, "Across the eighteenth century, evidence of the appeal of wax anatomy to female viewers is far stronger than that of their participation in the culture of dissection."[61] In particular, Rackstrow's Museum actively sought women to seek anatomical education. Clearly, the private anatomist collections entered the public sphere as forms of biological education, curiosity, and entertainment (not religious fervor). Waxworks of the famous and infamous circulated simultaneously with anatomical and pathological specimens of the criminal and the anonymous. Thereby, the display of parts became more palatable when objectified, anonymous, or infamous. Waxforms played a crucial role in this broader social acceptance of preserved human remains and spawned a cultural interest in all things corporeal.[62] Yet this Enlightenment objectification of the body also directly affected an attachment to and sentimentality toward the body. While the static body of the wax model served the anatomy student's empirical knowledge of the human form, wax mourning objects symbolize loss, status, and memory. Thereby, wax reshapes itself for more modern, individualized approaches to mourning and material attachments.

Undertaking and Mourning Wax

As we saw in Chapter 2, wax was integral to the modern embalming process. In the eighteenth century, wax cerecloths often kept a corpse from quick decay (and if full embalming could not be afforded, cerecloths would be the next best method). For both the anatomist and the undertaker, wax was an ingredient for preserving body parts and injecting veins and organs. Undertakers were often requested to paint a representation of the dead person's visage onto a cerecloth to reduce signs of putrefaction. In the case of the Hanoverian princess's embalming by John Hunter, we see the ingredient's significant role in the cerate: the corpse of George III's aunt Princess Amelia (d. 1786) was bound in green linen that had been steeped in a mixture of beeswax, resin, powdered verdigris, and mutton suet. Hunter biographer Wendy Moore notes that, "There were precise measurements for the lengths of the bandages required for each part of the body, and the method for waxing the fabric was similarly supplied."[63] Hunter explains, "Put the pack thread to the corners when you dip them and stand on a table to draw them easily out of the pan." Princess Amelia's body is then intricately swathed: "The body is wrapped up in two pieces & the face & head are covered with two pieces and afterwards rolled over with strips in every direction. The legs are then to be brought together and the two great toes tied & then all rolled up in one; the arms brought to the sides and the whole body is to be enveloped in two pieces each 7 feet long; the whole making one mass without any appearance of neck being retained."[64] This is an extreme example of preserving a royal, but the reliance on wax as a medium for the surgeon and undertaker is directly tied to the wider circulation of the dissected waxform. Better understanding of human biology and anatomy allowed for more precise anatomical models and improved preservation techniques.

A part of this sensibility for preserving the dead body is also reflected in the growing popularity of death masks. Most death masks are made by taking a cast of the face in wax or plaster, typically within hours after death. The history of death masks is rooted in ancient Rome, where wax death masks were often worn by actors as they mimicked the gestures of the deceased, and they have been found in ancient Egyptian tombs. Later, medieval tombs of royalty were noted to be adorned with the occupant's own death mask (and often including real hair). The practice then faded, but it eventually saw a revival in the eighteenth century.[65] An early example is the rare wax death mask of Charles Talbot, Earl of Shrewsbury (c. 1718) (fig. 3.4). Death masks were portable and reproducible, and they reflected the sensibility for handling and owning mementos that had infiltrated funereal practices. Sir Isaac Newton's mask, held by the Royal Society, is made of plaster with wax coating. The work of sculptor John Michael Rysbrack, it "was used in the preparation of Newton's memorial in Westminster Abbey."[66] Samuel Johnson's death mask was taken (by either sculptor James Hoskins or Joshua Reynolds) for the basis of a sculpture and subsequent artists' portraits.[67] Napoleon, the figure of egomania himself, requested that his entire corpse be wax molded. In a growing individualistic society, the self-interest in the desire to be remembered beyond the grave was of crucial importance, and the practice of having a death mask made lasted well into the nineteenth century. Before the advent of photography in the Victorian era, death masks were consequently one physical way by which to be remembered. Resembling embalming practices, the

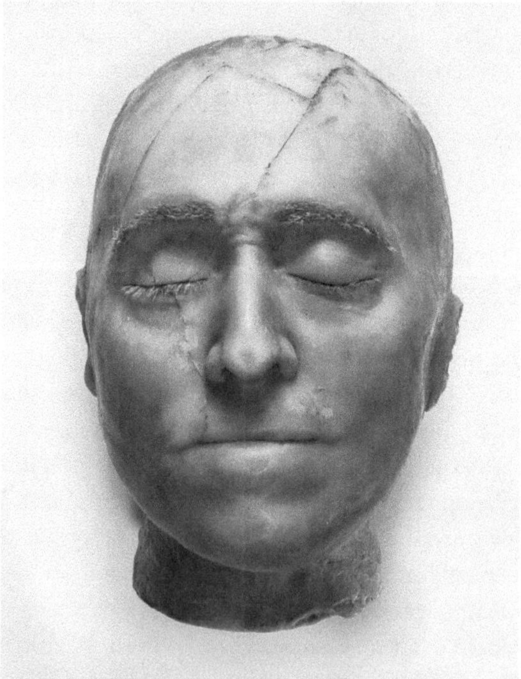

Figure 3.4. Death mask, Charles Talbot, Earl of Shrewsbury, c. 1718. Wax. © Fitzwilliam Museum, University of Cambridge.

death mask preserves parts of the corpse and therefore functions as a kind of auto-icon signifying a representation of its living counterpart. Romantics especially embraced the ancient form of symbolic immortality. A famous example is that of Keats, who on his death had his hand, foot, and face cast.[68] As with other forms of memorialization, death masks contain and engender an obsession with the corporeal and individualism.

Similar to death masks, wax bas-reliefs were a form of portraiture in the eighteenth century, and part of the trend in mourning practices that involved relics, such as the giving of mourning rings. Commissioned wax reliefs were especially popular among the aristocracy and royalty, and as a dimensional silhouette, they mimic the effect of the wax mask.[69] The middling class could afford these collectable mementos (which were much cheaper than Wedgwood jasperware reliefs) and would hang them on their walls. Many eighteenth-century examples commemorated former political leaders, such as Oliver Cromwell (who once again finds his head purchased and circulated), as well as war victories and heroes, such as the Duke of Wellington following the victory at Waterloo and Lord Nelson, modeled from life but produced for mourning purposes on his death. Low-relief portraits in wax were often exhibited at venues such as the Royal Academy and the Society of Artists.[70] Modeler Catherine Andras was awarded "The Larger Silver Pallet" by the Society for the Encouragement of Art, Manufacturers and Commerce for her memorial relief of Lord Nelson[71] (fig. 3.5). The likeness was much admired and was later used as the basis for the portrait by Conrad Heinrich Küchler in his design for the obverse of the Trafalgar medal, produced by Matthew Boulton. A contemporary wrote to Boulton in 1806, reporting a conversation with Lady Hamilton, Nelson's mistress, in which "she shewed me the inclosed Wax Profile which She declares is the most striking likeness that has been taken, & much more so than our little drawing or print by Mr Da Costa. On asking Lady H: in what features the model so closely resembled Lord Nelson as she had declared; she said, in the direction & form of the nose, mouth and chin, that the general carriage of the body was exactly his, and that altogether the likeness was so great it was impossible for anybody who had known him to doubt about or mistake it."[72] Though small, these wax reliefs could obviously be quite detailed and accurate, offering mourners a dimensional, flesh-like memorial of a loved one.

This ability to be molded into a dimensional form that stands in for the absent body is also reflected in the popular practice of fixing black mourning wax on letters, funeral invitations and tickets, wills, and other stationery, a practice that persisted into the nineteenth century. Wills or letters noting a dying person's

Figure 3.5. Catherine Andras, *Lord Nelson*, c. 1806.
Wax bas-relief memorial. © Victoria and Albert Museum, London.

last wishes were often legitimized by the placing of wax seals. Hallam and Hockey note, "Seals made of wax and imprinted with a sign associated with a particular person indicated property ownership but also attached traces of personal identity to the document. This was further reinforced when the seal was worn in the form of a ring with a shape or image cut into it . . . Locke away and protected with material enclosures, the will as materialized memory was again resonant with medieval metaphors of memory as a wax tablet or a strong box, but this time imbued with the stamp of the 'individual' and concealed within private, domestic spaces" (168). References to wax occur numerous times in *Clarissa*, notably the black wax used to seal letters from those in mourning. For example, in a letter to Colonel Morden from Clarissa, he notes that "the sealing of the cover was with black wax. I hope there is no new occasion in the family to give reason for black wax" (4:37, letter 7). Clarissa having preserved her epistles to be read after her death, we are told that Clarissa's packet of letters is "sealed up, with three black seals," and within are "eleven letters, each sealed with her own seal, and black wax" (8:15, letter 6). An invitation to Lord Nelson's funeral is decorated with a rare surviving seal (which would be made with a signet ring), impressed with his coat of arms and black wax (fig. 3.6).

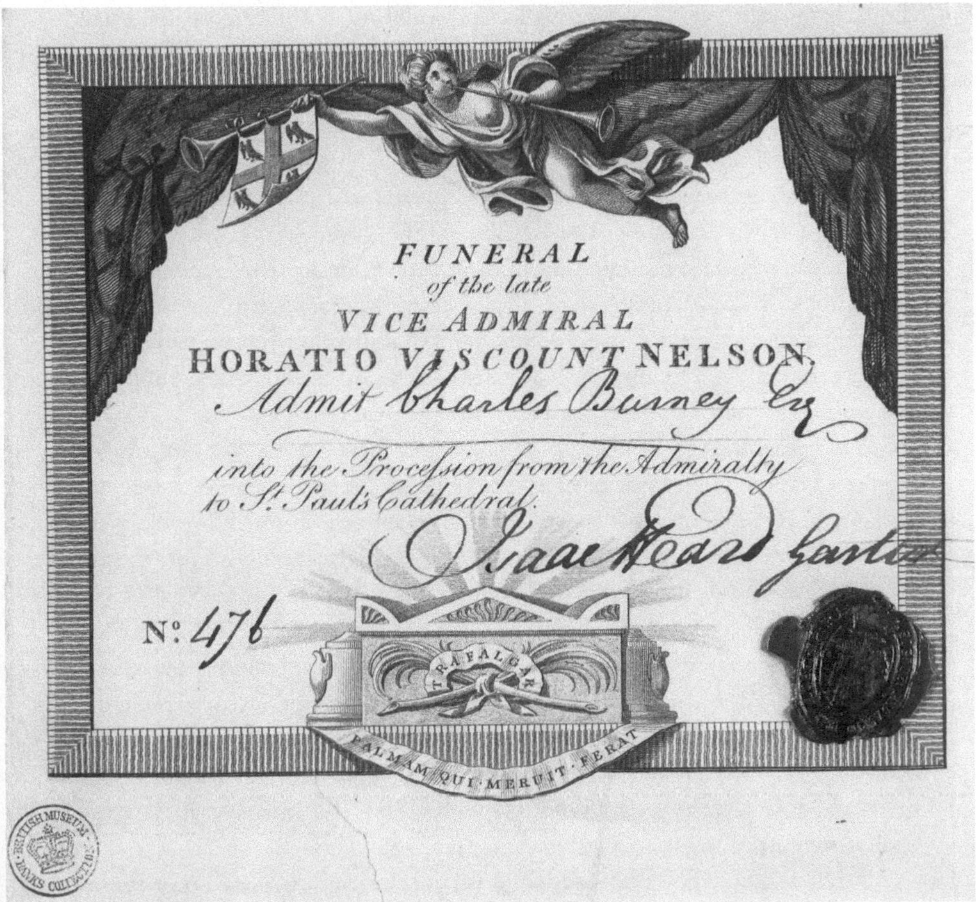

Figure 3.6. Ticket for the funeral of Lord Nelson with wax mourning seal, c. 1806. © The Trustees of the British Museum.

The origins of black-bordered stationery are seen in examples of seventeenth-century funeral cards, but the practice of black sealing wax emerged as a popular practice in the eighteenth century, as depicted in *Clarissa*. Some of the wax for sealing included aromatics, so there was a pleasant smell, and other formulas in the period ensured durability and included resin, turpentine, shellac, and pitch (and had an unpleasant odor). James Daybell explains in *The Material Letter in Early Modern England* that "The iconography of seals was often highly personalized and could hold symbolic meanings."[73] Black seals signified mourning, of course, and were produced by adding carbon, soot (lampblack), or dirt. During the early modern period, mourning letters would be routinely closed with the family seal pressed in black wax. In the eighteenth century, funeral invitations and tickets were part of the funereal equipage and were often presented with mourning seals, as in the example of Lord Nelson's invitation. In addition to broad black stationery borders, black seals not only visually signified the passing of another, but tactilely represented death. In Johann Zoffany's *Lady Salusbury in widow's weeds, with her spaniel "Belle"* (1765), painted four years after her husband's death, we see Lady Salusbury (mother of Hester Thrale) standing in mourning (frilled cap, pleated ruff, and black veil) before a portrait of her deceased husband with a document in her hand that probably represents her husband's will (fig. 3.7). Notable is the document's wax black seal, which physically and emblematically represents her husband's death and body.[74]

Associated with royal effigies, wax as an emblem of death and a medium for the preservation of the body has a lengthy history. Funerary statues were typically made of wood, but in the seventeenth and eighteenth centuries they were generally made of wax, and typically no longer carried at funerals.[75] Wax replicas of kings and queens were carried in funeral processions on top of coffins to "offset the effects of heat and putrefaction on the real bodies."[76] They provided a material representation of the physical body as well as a symbolic representation of the body politic. With this, effigies were intended to achieve the most striking realism possible. Formerly recumbent with hands crossed on top of the coffin, effigies in the eighteenth century include more examples of standing, lifelike waxworks. One example is that of William III (d. 1702), who is standing and dressed in full regalia (fig. 3.8).

When Edmund Sheffield, Second Duke of Buckingham, died on the course of the Grand Tour in October 1735, he was sealed in lead, and a replica made in wax, leather, wood, and fibrous stuffing bore his ceremonial costume at the state funeral in London.[77] The realism of this extraordinary visual object demonstrates its role as a replacement for the natural body (fig. 3.9). As a malleable, flesh-like medium, wax mourning relics were accessible to the middling and upper classes, and in its entertaining, anatomical, and effigy forms, wax circulated among all.

Described previously, the power of wax to replicate the fleshiness of the body makes it an appropriate choice for various forms of effigies and other *memento mori*. In a well-known instance, while in Florence, Sade visits the Palazzo Vecchio, and records his reaction to a case containing wax sculptures of corpses:

> One sees a sepulcher filled with an infinite number of cadavers, each one displaying the different gradations of decay, from a cadaver one day

Figure 3.7. Johann Zoffany, *Mrs. Salusbury in widow's weeds, with her spaniel "Belle,"* c. 1765. Oil on canvas. 50 × 40 in. © The Trustees of the Bowood Collection.

old to that which the worms have entirely devoured. This bizarre creation is the work of a Sicilian named Zummo [*sic*]. Everything is executed in wax and tinted in realistic colors. The impression is so overpowering that one's senses are utterly repelled. One instinctively covers one's nose without realizing it upon viewing this dreadful work, which it is difficult to see without bringing to mind the ghastly thoughts

Figure 3.8. Wax effigy, William III (d. 1702). © Dean and Chapter of Westminster, London.

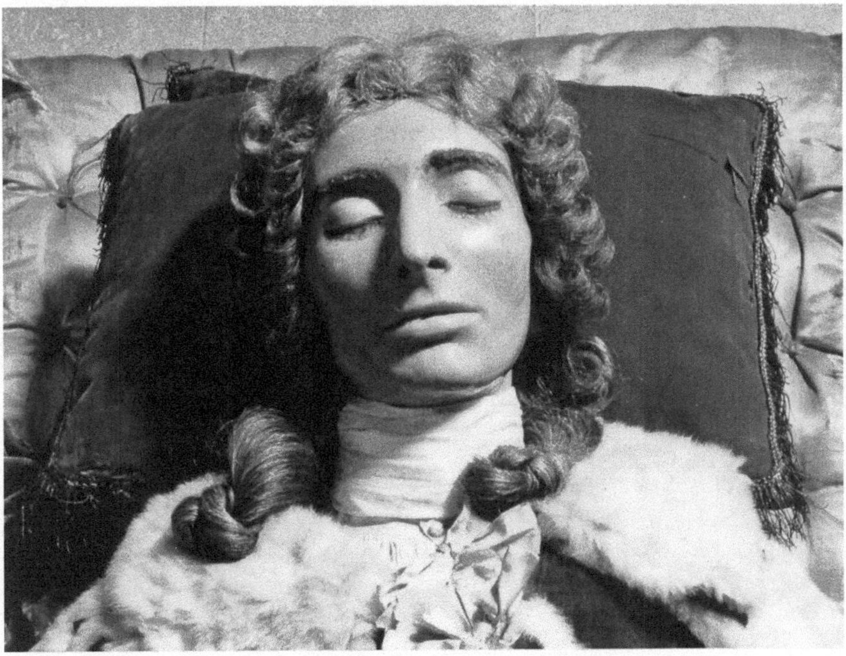

Figure 3.9. Wax effigy, Edmund Sheffield, Duke of Buckingham (d. 1721). © Dean and Chapter of Westminster, London.

of annihilation and, as a consequence, the more consoling thoughts of the Creator.[78]

The history of wax *memento mori* is long and is intimately linked with Catholic practices. In fact, post-Reformation Protestant critique of Catholicism included the rejection of wax reliefs of hell, wax statues of saints, and other forms of religious aesthetics. As Sade describes (and later inserts his impression in *Juliette*), Zumbo's theaters of death, his tableaux of highly expressive wax figures arranged and depicted in various stages of decay, forced viewers to question their sins and faith. For example, "The five stages of decomposition" was modeled in wax relief and depicted grisly scenes of death and decay inspiring moral lessons.[79] The wax theater depicting Voltaire's death is a notable example of this (fig. 3.10). Reminiscent of Tussaud's wax model of the death of Marat, in Voltaire's *théâtre de la mort*, sculptor Christopher Curtius emphasizes the magnitude of the cultural icon (who was deist) in the painful throes of death. As Thomas Laqueur remarks of the relief, "At the bedside, the negotiations were about what exactly Voltaire

Figure 3.10. Christopher Curtius, *The Death of Voltaire*, c. 1760–1795. Colored wax. Paris. © Victoria and Albert Museum, London.

would have to do for the Church, whose most famous enemy he had been, to grant him a proper burial."[80] Yet the theaters of death in Radcliffe's fiction steer readers away from this religious purpose.[81] As we have seen, wax models supplied counterfeit versions of death. In their malleable, life-like forms, they also operate as anti–*memento mori*. Symbolizing the tension between the unreal and the real body, inanimate renderings of flesh invite and repel touch. This uncanny aspect of waxworks (which we find in life-like dolls, puppets, effigies, and so forth) participates in reminding viewers of their mortality while simultaneously reinforcing one's liveliness by sensory engagement (seeing, touching, or smelling the life-like/less wax).

Radcliffe's Wax

As wax materials increased in popularity, and were a part of cultural discourse, wax items saw circulation in the period's literature, especially in the novel. Novels are receptacles of materiality, and when intersected with death and mourning, they provide a space for objects to embody their own mysteries and plots. I wish to close this chapter with a discussion of the role of wax in Ann Radcliffe's *The Mysteries of Udolpho* in order to demonstrate the potential for objects to take on such narrative significance. In particular, the multifaceted and dynamic materialism of wax—philosophical, entertaining, anatomical, religious, and sentimental—is transformed and reshaped by Radcliffe to ultimately signify her Protestant worldview informed by Enlightened thought and reason. Radcliffe's placement of her Gothic novels in traditionally Catholic, historical settings helps her English readers locate remnants of Catholicism in France and Italy. Thus patriarchal tyranny, papist iconography, and nationalism are displaced, and her novels ultimately reinforce Protestant Englishness. Invoking Catholic iconography and allowing her mostly Catholic protagonists to indulge in superstition and the supernatural (charges made against papism in the period), her novels then violently expurgate the supernatural. Wax symbolizes this tension and, in my reading, launches Emily St. Aubert's secular individualism and rational, empirical lesson in *Udolpho*.

The epigraph to this chapter from *Northanger Abbey* directly refers to the Gothic plotting of Radcliffe's *A Sicilian Romance* (1790): Julia's mother the marchioness is imprisoned in a dungeon, a waxen effigy of her is indeed introduced, and a "supposititious funeral" is carried out. In Radcliffe's plot, the married Marquis of Mazzini desires to marry a younger woman, Maria de Vellorno, thereby he incarcerates the marchioness in an underground vault for the rest of her life, claims she is dead, and plans her funeral, all in order to marry Maria. Toward the end of the novel, when Julia discovers her mother alive in the vault, the marchioness explains her lengthy incarceration and false funeral: "'I afterwards learned, that in obedience to the marquis's order, I had been carried to this spot by Vincent during the night, and that I had been buried in effigy at a neighboring church, with all the pomp of funeral honor due my rank.'"[82] We are told that the marquis indeed buried a wax effigy of her, a device that will recur a few years later when a strange wax figure haunts *The Mysteries of Udolpho*. Diane Long Hoeveler remarks, "Although the mother was not literally buried alive, she might as well have been, for she tells Julia that each day passed 'in a dead uniformity, more dreadful than the most acute vicissitudes of misfortune'" (177).[83] The dead-mother plot is centered on the effigy and the ability for wax to stand in for the dead body. "The pomp of funeral honor" and wax performance satisfy observers: no one suspects any wrongdoing.

This strange incident is fully developed in *The Mysteries of Udolpho, A Romance* (1794), in which Radcliffe centers the novel's mystery on a waxwork that stands in for the dead body. Courtney Wennerstrom's feminist argument reads Radcliffe's work as the "subversive evocation of anatomy's deeply ingrained connections to criminality, necrophilia, sadism, voyeurism, and pornography," enabling her to "slice through male control over female eroticism and maternity"[84] (196). Her reading of Radcliffe's protofeminist intention is too generous, but it does assist my larger point that the dead body and its replications in her fiction have the ability to empower the female ingénue. While Wennerstrom, Bronfen, and others read the Anatomical Venus and other human effigies as visualizations of exposed, invasive female sexuality for male ocular control and consumption, I read Emily's exposure to the dead body as self-affirming. For her enlightened transformation, Emily needs to overcome her sensibility through the reworking of the ocular message of the dead body (not just the female corpse).

To support this claim, it is necessary to quote in full the well-known passage describing Emily's encounter with the mystery behind the black veil:

> To withdraw her thoughts, however, from the subject of her misfortunes, she attempted to read, but her attention wandered from the page, and, at length, she threw aside the book, and determined to explore the adjoining chambers of the castle. Her imagination was pleased with the view of ancient grandeur, and an emotion of melancholy awe awakened all its powers, as she walked through rooms, obscure and desolate, where no footsteps had passed probably for many years, and remembered the strange history of the former possessor of the edifice. This brought to her recollection the veiled picture, which had attracted her curiosity, on the preceding night, and she resolved to examine it. As she passed through the chambers, that led to this, she found herself somewhat agitated; its connection with the late lady of the castle, and the conversation of Annette, together with the circumstance of the veil, throwing a mystery over the subject, that excited a faint degree of terror. But a terror of this nature, as it occupies and expands the mind, and elevates it to high expectation, is purely sublime, and leads us, by a kind of fascination, to seek even the object, from which we appear to shrink.
>
> Emily passed on with faltering steps, and having paused a moment at the door, before she attempted to open it, she then hastily entered the chamber, and went towards the picture, which appeared to be enclosed in a frame of uncommon size, that hung in a dark part of the room. She paused again, and then, with a timid hand, lifted the veil; but instantly let it fall—perceiving that what it had concealed was no picture, and, before she could leave the chamber, she dropped senseless on the floor.[85]

Emily's experience is filled with curiosity, excitement, and terror; she seeks out the object she fears. Overwhelmed with the terrifyingly sublime mystery, she inevitably faints, losing her sensory capacity. Though Emily seems to fetishize the pleasurable terror, it is the process of regaining her senses and reason that Radcliffe reinforces.

First described as a "picture" and then "no picture," the object behind the veil remains a mystery, and the reader is clueless as to what has exactly prompted Emily to faint. Radcliffe continues:

> When she recovered her recollection, the remembrance of what she had seen had nearly deprived her of it a second time. She had scarcely strength to remove from the room, and regain her own; and, when arrived there, wanted courage to remain alone. Horror occupied her mind, and excluded, for a time, all sense of past, and dread of future misfortune: she seated herself near the casement, because from thence she heard voices, though distant, on the terrace, and might see people pass, and these, trifling as they were, were reviving circumstances. When her spirits had recovered their tone, she considered, whether she should mention what she had seen to Madame Montoni, and various and important motives urged her to do so, among which the least was the hope of the relief, which an overburdened mind finds in speaking of the subject of its interest. But she was aware of the terrible consequences, which such a communication might lead to; and, dreading the indiscretion of her aunt, at length, endeavoured to arm herself with resolution to observe a profound silence, on the subject. (248–49)

In this recovery scene, Radcliffe's syntax points to the maturation Emily will undergo over the course of the narrative: "recovered," "strength," "regain," "courage," "reviving," "hope," "arm herself with resolution."

In order to overcome this trauma and superstition, and be prepared for the recognition and sensory acknowledgement of the veiled mystery, Emily must relive the horror over the course of hundreds of pages. In fact, the word "veil" is used seventy-three times in the novel, and at least eight of these directly refer to Emily's recollection of the veiled rotting corpse. For example: "The horror of the chamber rushed on her mind. Several times the colour faded from her cheeks, and she feared, that illness would betray her emotions, and compel her to leave the room; but the strength of her resolution remedied the weakness of her frame; she obliged herself to converse, and even tried to look cheerful" (249). Again, though she has a somatic response to traumatic revisions of the mysterious horror, phrases such as "the strength of her resolution" signal her overcoming of emotional weakness. As if describing a house of wax or Cox's Museum of automata, the narrator uses the word "spectacle" nineteen times in reference to the horrific mystery, such as "the veil had formerly disclosed a dreadful spectacle" (320) and "As she returned towards her chamber, Emily began to fear, that she might again lose herself in the intricacies of the castle, and again be shocked by some mysterious spectacle" (252). Thereby, Radcliffe instructs her heroine (and titillates her readers) through repetitive reminders of the shock and awe: "As she passed the door of the apartment, where she had once dared to lift the veil, which discovered to her a spectacle so horrible, that she had never after remembered it, but with emotions of indescribable awe, this remembrance suddenly recurred. It now brought with it reflections more terrible, than it had yet done, which the late conduct of Montoni occasioned; and, hastening to quit the gallery, while she had power to do so, she heard a sudden step behind her . . . and all the horrors of that chamber rushed upon her mind" (384). Through these repetitions of "the spectacle she had witnessed in a chamber of Udolpho . . . the horrible appearance, disclosed by the black veil" (491), Emily eventually overcomes her fears and superstitions. The pleasurable horror loosens its grasp on her imagination by the time Radcliffe unveils the mystery, and Emily's sensory capacity is strengthened.

The revelation that what Emily St. Aubert thought was the decaying, embalmed corpse of the murdered Signora Laurentini was actually a waxwork *memento mori* is generally read as anticlimactic. Emily reads the veiled portrait as the mutilated corpse of Laurentini largely because she is piecing together partial bits of information that do not cohere around any other reading (Hoeveler 96). Emily desires it to be an actual corpse, and she projects this desire onto the waxen form. We have seen her morbid drive to unveil the corpse (literally and imaginatively). But in fact by the end of the book we learn that quite another reading is the "correct" one for this wax figure: it is a religious icon constructed by the monk of the abbey, meant to be a meditation on the inevitably of death. For Emily, the waxwork has a different effect than Zumbo's *memento mori*: her traumatic reaction is not one of religious supplication or the acceptance of mortality. As Terry Castle reminds us, "Christian icons serve a largely decorative or aesthetic function in the novel," and the "numerous monks and nuns who pop up in the novel function more as picturesque adjuncts to the action than as emblems of religious authority."[86] Though Castle reads the waxwork as an emblem of sentimental and erotic reveries, I wish to read its symbology with the Enlightenment context this chapter has charted. Shockingly realistic, the waxwork serves as a reminder of superstition and weakness throughout the novel; brief reminders invoke thoughts of fainting horror. These traumatic touchstones ultimately bolster Emily's fortitude. Wax thereby melds into a different purpose: it symbolizes her conquering of the imagination. First, the narrator describes Emily's active imagination and the horrid details of the waxwork:

> Of a crime, however, to which Emily had suspected, from her phrensied confession of murder, that she had been instrumental in the castle

> of Udolpho, Laurentini was innocent; and she had herself been deceived, concerning the spectacle, that formerly occasioned her so much terror, and had since compelled her, for a while, to attribute the horrors of the nun to a consciousness of a murder, committed in that castle.
>
> It may be remembered, that, in a chamber of Udolpho, hung a black veil, whose singular situation had excited Emily's curiosity, and which afterwards disclosed an object, that had overwhelmed her with horror; for, on lifting it, there appeared, instead of the picture she had expected, within a recess of the wall, a human figure of ghastly paleness, stretched at its length, and dressed in the habiliments of the grave. What added to the horror of the spectacle, was, that the face appeared partly decayed and disfigured by worms, which were visible on the features and hands. (663)

Next, the narrator reinforces the origins and religious intention of the waxwork:

> On such an object, it will be readily believed, that no person could endure to look twice. Emily, it may be recollected, had, after the first glance, let the veil drop, and her terror had prevented her from ever after provoking a renewal of such suffering, as she had then experienced. Had she dared to look again, her delusion and her fears would have vanished together, and she would have perceived, that the figure before her was not human, but formed of wax. The history of it is somewhat extraordinary, though not without example in the records of that fierce severity, which monkish superstition has sometimes inflicted on mankind. A member of the house of Udolpho, having committed some offence against the prerogative of the church, had been condemned to the penance of contemplating, during certain hours of the day, a waxen image, made to resemble a human body in the state, to which it is reduced after death. This penance, serving as a memento of the condition at which he must himself arrive, had been designed to reprove the pride of the Marquis of Udolpho, which had formerly so much exasperated that of the Romish church; and he had not only superstitiously observed this penance himself, which, he had believed, was to obtain a pardon for all his sins, but had made it a condition in his will, that his descendants should preserve the image, on pain of forfeiting to the church a certain part of his domain, that they also might profit by the humiliating moral

> it conveyed. The figure, therefore, had been suffered to retain its station in the wall of the chamber, but his descendants excused themselves from observing the penance, to which he had been enjoined. (663)

Though Emily imagined the mutilated figure was the body of Signora Laurentini, allegedly Montoni's former lover, we learn that it is a theatrical prop that signifies no particular person (though some read it as a wax image of the Marquis of Udolpho's corpse). As with Zumbo's theaters of death, the "face appeared partly decayed and disfigured by worms, which were visible on the features and hands." The waxwork is to serve as a reminder of the body's impermanence and to inspire viewers to question the state of their souls. "Romish," "superstitiously," "pardon for all sins," "condemned to the penance," "humiliating moral" all designate the waxwork's Catholic message. Despite the marquis' direction in his will, his descendants do not observe the "penance" and keep it hidden from view, thus it is rendered powerless as a religious icon. And Emily's unveiling of the waxwork seems to reinforce that the icon is devoid of moral power. Ann Kibbie remarks that "Emily's own visceral response has already served as a rejection of the icon's original, didactic intent," and that she is alienated from the *memento mori* tradition of contemplation.[87] Her somatic response is similar to Sade's when he enters La Specola: the rotting corpses are so lifelike he quickly covers his nose with his hand, fearing their stench, while *all* of Emily's senses shut down. Throughout the novel she reminds herself of its unspeakable horror but in no way does she experience religious humility.

Now that the religious spectacle has been fully explained, the narrator must remind the reader that Emily should not be faulted for her somewhat extreme and excessive sensibility: "This image was so horribly natural, that it is not surprising Emily should have mistaken it for the object it resembled, nor, since she had heard such an extraordinary account, concerning the disappearing of the late lady of the castle, and had such experience of the character of Montoni, that she should have believed this to be the murdered body of the lady Laurentini, and that he had been the contriver of her death" (663). Not able to fully process what she had seen, Emily was "compelled" to believe "that Montoni, not daring to confide the secret of [Laurentini's] death to any person, had suffered her remains to decay in this obscure chamber."[88] Emily's belief that an actual decaying corpse could be "housed" and kept (like Clarissa's) is striking. Yet this is veiled in mystery for hundreds of the novel's pages. The narrator explains that "it was the dread of his terrible vengeance, that had sealed her lips in silence, concerning what she had seen in the west chamber" (662–63).

It is this persistent silence, coupled with reason and skepticism, that successfully projects Emily through numerous other suspicious, supernatural circumstances. Though never devoid of superstition, Emily learns to gradually control her sensibility and spectral imagination. For instance, when she hears a mysterious sound, she is now able to talk herself through her active imagination: "She paused, for, notwithstanding the terrors of darkness, she considered it to be very improbable, whoever he was, that he could have any interest in intruding upon her retirement; and again the subject of her emotion changed, when, remembering her nearness to the chamber, where the veil had formerly disclosed a dreadful spectacle, she doubted whether some passage might not communicate between it and the insecure door of the stair-case" (320). "Very improbable" and "doubted" signal her more reasonable reaction. Later, when she is confronted with an actual corpse (which she also misidentifies as a woman—in this instance, her aunt), she once again faints, as she did on viewing the wax portrait. Yet in this instance she immediately describes the corpse, and we learn it is a wounded but not as yet decomposing dead soldier. She does not replay the image or horror, but describes "the features deformed by death" and that the "livid wound appeared in the face" (348). This response to graphic death shows maturation and growth catalyzed by her sensory acceptance of the permanently decomposing wax portrait.

At the end of the novel, Emily gives a speech that should convince the novel's readers that the supernatural (for the most part) has been exorcised from her imagination: "'I perceive,' said Emily, smiling, 'that all old mansions are haunted; I am lately come from a place of wonders; but unluckily, since I left it, I have heard almost all of them explained'" (491). This pronouncement is quite a shift from Emily's previous mode, as the narrator reminds us: "Emily felt herself still inclined to believe more of the wonderful, than she chose to acknowledge" (491). In typical Radcliffean fashion the narrator explains "all" of the "wonders"; Emily and the reader can rest peacefully now that the ghosts and spectacles have been explained away and buried. This purging of what Radcliffe's readers would consider "Romish" superstition inevitably results in the reinforcing of Protestant, English values. Traumatized yet empowered by the mystery behind the veil, Emily learns to ultimately overcome superstition and control her rampant imagination. In one example at the end of Volume III, Emily once again hears mysterious lute music that had previously frightened her. Surrounded by those who still believe the music emanates from a "spirit," Emily's reaction is finally more reasonable. The narrator tells us "Emily smiled, and remembering how lately she had suffered herself to be led away by superstition, determined now to resist its contagion" (490). By Volume IV, Emily is blushing with embarrassment when others explain what she formerly believed to be actions of ghosts, such as the hidden banditto, who, terrified to be caught, moves the bedclothes in Emily's room and feigns he is a spirit. Emily's maturation is rewarded with an inheritance and management of her father's estate by the end of the novel, which speaks to the larger themes of the health of the English economy and rightful inheritance and birthright. The narrator reinforces this moral by pronouncing the happiness of the rejoined young lovers, Emily and Valancourt, restoring "the securest felicity of this life, that of aspiring to moral and laboring for intellectual improvement—to the pleasures of enlightened society, and to the exercise of benevolence" (672). With this, Radcliffe implies that the greater good of English society and its betterment are centered on rational Enlightenment values.

This chapter has argued that as a fleshly form of relic, variations of wax and its circulation—as iconography, *memento mori*, mourning seal, waxwork, anatomical model—register the growing cultural shift for personal, tangible engagement with the body and corpse. An ancient material used for effigies, wax is modernized for eighteenth-century mourning purposes: bas-reliefs, seals, and waxworks materialize loss and are talismans of memory. Flesh, the medium of eighteenth-century knowledge, quickly decomposes. Anatomists thereby created wet and dry specimens, waxforms, and atlases of the dissected body in order to preserve the knowledge they cut into flesh. The translucent, sweaty resemblance to resisting flesh works as a perfect medium for not only circulating houses of waxworks, but for the anatomist, wax modeler, and undertaker. And as the chosen element for Catholic *memento mori*, wax symbolizes the intersection of the religious and scientific. As icons and religious relics were purged from a post-Reformation, Protestant England, the culture seemed to embrace a secular wax medium as a desire for the thingness of the body was popularized. This veneration of remains directly correlates with Radcliffe's waxwork. Emily's supernatural exorcism and moral education are ultimately facilitated by *Udolpho*'s wax *memento mori*, yet they are devoid of Catholic iconography. For Emily, the decomposing wax figure does not remind her of her mortality, the need to confess, or the state of her soul. Radcliffe's own Zumbo-like *ars moriendi* is enlightened materialism in all its decomposing, sweating, horrific glory.

CHAPTER 4

Circulating Bodies

Secular Mementos, Jewelry, and Hairwork

When my grave is broke up againe
Some second ghest to entertaine,
(For graves have learn'd that woman head,
To be to more then one a Bed)
And he that digs it, spies
A bracelet of bright haire about the bone,
Will he not let'us alone,
And thinke that there a loving couple lies,
Who thought that this device might be some way
To make their soules, at the last busie day,
Meet at this grave, and make a little stay?

—John Donne, "The Relique"[1]

When notified of Clarissa Harlowe's death, her rapist Robert Lovelace insists that she be dissected, parceled, and preserved: he requests her heart and viscera. In addition, he requests a common eighteenth-century memento: hair. "And a lock of her hair I will have, let who will be the gainsayers!" (8:44, letter 17).[2] Lovelace never receives her organs or the lock of her hair; in addition, as Kathleen M. Oliver reminds us, "No member of the Harlowe family will receive a lock of her hair, in any form."[3] Numerous friends and relatives receive rings without her hair. Clarissa's maternal aunt is bequeathed "the sum of Fifty guineas for a ring" and her cousin Dolly "Twenty-five guineas for a ring, to be worn in remembrance of her true friend" (8:102, letter 27). In her will, Clarissa also directs that "four charming ringlets" be cut from her hair and encased in mourning jewelry ("with my hair in crystal"), an intimate and costly gift bequeathed to a chosen few (8:105, letter 27), including Anna Howe, Anna's fiancé Mr. Hickman, as well as members of Lovelace's family. Colonel Morden claims one of Clarissa's tresses "for a locket, which, he says, he will cause to be made, and wear next his heart in memory of his beloved cousin" (8:41, letter 15). Belford relates to Lovelace, "As she has directed Rings by her Will to several persons, with her hair to be set in crystal, the afflicted Mrs. Norton cut off, before the coffin was closed, four charming ringlets" (8:41, letter 15). A fifth and final ringlet is cut after Anna Howe pleads that she "might be allowed a lock of the dear creature's hair" (8:81, letter 22). In her will, Clarissa underscores the importance of preserving tokens of her hair: "I make it my earnest request to my dear Miss Howe, that she will not put herself into mourning for me. But I desire her acceptance of a ring with my hair; and that Mr. Hickman will also accept of the like; each of the value of fifteen guineas." Several relatives receive very intricately detailed mourning rings: "an enamelled ring, with a cypher CL. H. with my hair in crystal, and round the inside of each, the day, month, and year of my death: Each ring, with brilliants, to cost twenty guineas" (8:105, letter 27). One receiver of the mourning ring, Lovelace's cousin Charlotte Montague, remarks the rings "will be long, long worn in memory" (8:119, letter 30). As Richardson's *Clarissa* intimately details, by mid-century the Church's control over funeral practices was loosening and the undertaker's tasks were increasing in importance. Highly personalized, Clarissa's funeral is planned by herself, reflecting the rejection of established rituals for a more individual approach. In addition to having an undertaker (who solicited jewelers and hairwork artists) produce mourning lockets and rings with her hair in crystal, Clarissa designs and paints her own coffin, writes her own will, and, as she directs in her letters and will, makes sure she is not dissected, a practice, as previously discussed, not uncommon for the wealthy.[4] Her requests are intimate and individualized, aesthetic and self-interested.[5] The dissemination and preservation of her hair symbolizes her own form of embalming: she directs which "parts" of her body are intricately preserved and treasured for mourners. While Lovelace desired to have control over her corpse and its preservation, Clarissa ensures posthumous control through her body's transformation into circulating, secular relic.

As I have argued in the previous chapters, the concern over the dead body—its preservation, parceling, and display—takes on renewed interest and concern in the period, and we see this dramatically play out in fiction. In their examination of hair and bone, these last two chapters claim that these changes resulted in a sensibility toward the dead body and a cherishing of its parts. These practices were sentimental, intimate, and personalized, while also commodified, class-oriented, and public. Yet the rise in popularity of hair jewelry especially in the second half of the century does underscore a sentimental turn; memories and emotional attachments to the dead are bound in these mementos. Hallam and Hockey aptly describe how mortuary sculpture attempted to capture this intersection of bodily dissolution and permanence:

> In early modern England, for instance, the prevailing Christian notions about death as the release of the spiritual from the material body defined the corpse as matter that would fade away, in contrast to the continuity of the spirit that lived on. In this setting, the corpse was buried, while the social presence of the body could be maintained via a stone or marble funerary monument that provided a sculpted body as in life. The corpse after disposal was allowed to recede, to return to dust, while the stone or marble image of the body, made visible for purposed of memory, was one that marked a death while referring back to a "life"—the personhood of the deceased was thus partially retained when visual signs of social belonging and status were deployed by, for example, displaying the deceased in sculpted clothing and jewellery. Funerary monuments of this type rest upon a precarious boundary somewhere between the status of a corpse and a socially "living" body. (132)

This preservation of "personhood" sees a reliance on the material body in the eighteenth century. Hallam and Hockey note this significant cultural shift: "Further memory practices have sought not to replicate the corpse, but to salvage and deploy it directly. Rather than using materials such as stone, marble or ivory, to represent the body, it is the flesh itself, or bodily substances such as bone, blood and hair that are regarded as powerful memory 'objects'" (134). These secular relics thereby operate as *persons* but also *things*. As we have seen in earlier chapters, from the early modern period on, various forms of secular relic veneration increased as the faith in saints and their relics declined. In the context of his discussion of William Godwin's "Essay on Sepulchres" (1809), Thomas Laqueur remarks, "One can go a long way toward making the dead body consequential without having particular eschatological or metaphysical commitments" (54). As both the dying and mourners alike were concerned about bodily dissolution, mementos saw growing importance and incorporated hair and bone from the dead and/or the mourner.

Inanimate objects, like people, are agents that shape social formations and interactions. Objects embody histories and play important roles within the experience of grief and the expression of mourning. In *Relics of Death in Victorian Literature and Culture*, Deborah Lutz has convincingly argued for a Victorian culture that treasured objects that belonged to the dead, and for a literature that singled out the keepsake and thingness of personal objects. Her definition of "secular relics" is integral to my study. Though she does not fully discuss this, she sees the emergence of the "relic collecting of nonsaints" at the end of the eighteenth century, and identifies that this "was not, in the main, a religious practice."[6] Lutz recognizes that while some forms of religious belief bolstered the cherishing of remains (such as Victorian Evangelicalism, Spiritualism) "the ritual had largely pagan motives, especially for the Romantics" (21). I agree with her assessment, but as my study shows, these practices fully emerge early in the eighteenth century as Enlightenment culture navigated faith and religion along with scientific discovery. While Lutz rightly observes that the desire for secular body parts and other materials of death "had its roots in the longings for the relics of the saints and other venerated figures" (21), I wish to trouble this traditional narrative by exploring eighteenth-century relic culture in a post-Reformation, Enlightenment context. Peter Marshall states, "Despite the dissolution of chantries and fraternities, the proscription of requiem masses and intercessory prayers, the putting out of obit lamps, and abrogation of bede-rolls, Elizabethan and early Stuart England possessed a plethora of methods and occasions for memorializing the dead, and sanctifying their memory."[7] Not all hallowed remains were destroyed in the Reformation, and many survived undisturbed in cathedrals and minsters (like Westminster and York).[8] But as we will witness in this and the next chapter, various forms of secular relic veneration increased as the faith in saints and the power of their relics declined.[9]

Though for some secular relic culture may be seen to mirror Catholic practices, the Protestant campaign to "discredit the cult of relics" (part of Cromwell's war against the Church of Rome) actually produced a reinvention or refashioning of the practice largely without pious intent.[10] Objects, emblems, and tokens of memory became part of a more modern, secular identity formation. (Even Gibbon mentions in the *Decline and Fall* [1781], "The worship of saints and relics corrupted the pure and perfect simplicity of the Christian model").[11] So while the resurgence of forms of behavior toward human remains seemed to be outwardly reminiscent of the discredited cult of relics, and, in Alexandra Walsham's terms, "attracted charges of Protestant hypocrisy" (143), we should also acknowledge that in many instances the practice remained purely commemorative.

Curator and historian Marcia Pointon remarks that "Protestants in this period were both fascinated and repelled by relics they saw on their travels, yet they thought nothing of imprisoning locks of hair in an artifact that might be worn as a kind of talisman."[12] Increasingly industrialized and commodified, religiously pluralistic and diverse, Protestant culture experienced memory and identity in forms that were inevitably bound to involve the ritualization of material objects. This imbrication between the sacred and souvenir, the supernatural and artwork was to be anticipated. Stewart explains, "Because they are souvenirs of death, the relic, the hunting trophy and the scalp are at the same time the most intensely potential souvenirs and the most potent antisouvenirs. They mark the horrible transformation of meaning into materiality more than they mark, as other souvenirs do, the transformation of materiality into meaning" (140). While the return of saints' relics can offer a form of restoration, for example, souvenirs offer an end to this sacred narrative. Walsham highlights that this diminishing religious importance is signified by the fact that "ancient Catholic relics seem to have gradually migrated into the realm of historical artefact" (140), and that many holy objects and relics "ended up in contemporary cabinets of curiosities."[13] As examples, she notes that pieces of John the Baptist's tomb and one of the thorns Christ was crowned with were displayed with Henry VIII's stirrups and Anne Boleyn's embroidered night veil (141). Even the word "relic" was losing its explicitly Catholic connotations in the period, more often denoting things kept for the remembrance of persons, places, periods, and even remnants of past belief.[14] For post-Reformation Protestants, the place of the dead had to be reviewed and renegotiated in light of such a momentous social and theological realignment. The outlawing of prayers and masses for the dead, with the dissolution of the concept of purgatory, "virtually redefined the relationship of the living to the dead and resulted in a widespread sense of powerlessness on the part of the bereaved to influence the fate of their departed and anxiety about the prospect of facing the suddenly final sentence of the deathbed."[15] Isolated from the dead, the living sought new avenues for channeling their grief and new forms of commemoration. As material markers for these large changes, secular memorials eased the discomfort of a violently reformed religiosity.

While we see this aesthetic emerge a bit later in America, British mourning culture fully accepts representations of physical contact between the living and the dead early in the period.[16] Decay inevitable, small relics of the body thereby serve as morbid substitutes for the dead. These funerary accoutrements function as secular, tangible, wearable reliquaries that house corporeal mementos (or symbols for them). Their prominence is seen in numerous wills and letters from the period, signifying the sentimental and aesthetic approach to mourning. Post-Reformation England saw a shift from human remains receiving pious veneration to remains becoming objects of scientific scrutiny and public display. As previously argued, I see funerary customs influenced by the growing circulation of human specimens in the period. And as dissection was integral to Enlightenment anatomy, individual control over the corpse was of growing concern. Thus, the parceling and preservation of the dead body stemmed from fears of dissection and dissolution, sentimentalism, and an increased doubt about the afterlife.

Constance Classen discusses the "cult of relics" during the Middle Ages in *The Deepest Sense: A Cultural History of Touch* and notes the difference between the saintly and the lay corpse: "Ordinary corpses might decompose and stink, but saintly bodies were imagined to be incorruptible and fragrant. The relic of a saint, hence, was understood to be quite a different thing from handling ordinary human remains."[17] Saintly relics—which were handled, worn on the body, kissed, and even put into the mouth—embodied a supernatural force "that had the power to grant one good health, good fortune, and a good end—all through the medium of touch" (Classen 40). Relics were believed to protect from harm those who were in contact with them. "Medieval travelers, hence, often carried some small relic on their persons, often in a bag strung around the neck," Classen explains (38–39). Knights would enclose relics in their swords, believing the saint could supernaturally grasp the sword for protection in battle.[18] In the eighteenth century, the handling of human remains is renegotiated; lay bodies are sentimentalized, are commodified, and receive commemorative devotion. They are not conduits to the supernatural or containers of God's power. Sean Silver recognizes this shift as such: "Though seventeenth- and eighteenth-century England countenanced few official relics, numerous other charged objects centered similar ritual processes," such as "the clothes and insignia of Charles I—the so-called martyr-king—were especially charged objects, rising to the status of relics among his followers" (175). Following his execution, his blood and hair are distributed and transformed into personal objects, "in an enameled locket, a blood-dyed book cover, and numerous miniatures worked in needlepoint from his hair."[19] For my discussion, it is the British Enlightenment's transformation of body parts into sentimental, devotional objects that signals this new cultural direction. Personal property of the dead, as well as objects made from the dead, embody a personal potency—they are mementos of love and friendship and signify individual memorialization. Hair and bone (such as Milton's), embalmed viscera and hearts (such as Clarissa's), kidney and bladder stones (such as William Hay's and Samuel Pepys's), teeth, and even nail clippings, become what Sean

Silver terms "hypercharged" relics (175). Simultaneously, body parts become collectible and fashionable, devoid of love and personal devotion. Fueled by the growing collections of anatomists, they are transformed into souvenirs of the famous and infamous and are often aestheticized. For example, Horace Walpole owned several locks of hair, including that of Mary Tudor, queen of France, housed in a locket.[20] This chapter will interrogate these practices, demonstrating that Clarissa's desire for her bodily remains to be deployed as jewelry is far from being disturbing or culturally unacceptable.

The giving of memorial and mourning rings is an ancient custom. The earliest mourning rings were somber, often with religious symbology: Christ on the Cross, the five wounds, the nails and the whip. While these common artifacts strike modern sensibilities as morbid, they have precedents in *memento mori* jewels of the medieval period and were popularized in the Renaissance. Along with Sean Silver, Ginny Dawes identifies that mourning jewelry sees a resurgence following the death of Charles I: "The execution of King Charles I in 1649 spawned a jewellery industry which produced carved rock crystal stones that were mounted over ciphers made of gold wire, displayed upon a background of hair work, and set in simple old rings to show support for the monarchy."[21] It is from these roots that the popularity of the eighteenth-century memorial ring began to grow. Seventeenth-century English mourning slides were frequently worn as pendants, "depicting such motifs as skulls with cross-bones or skeletons holding hourglasses" (Frank 119). Early mourning rings with these *vanitas* symbols brought comfort and were given to the bereaved at funerals and bequeathed from generation to generation. Rings were often designated in wills, and the amount of money to be spent on purchasing them was regularly quoted. Famed actress Nell Gwyn (d. 1687) left fifty pounds to Lady Fairbourne to buy herself a ring "in memory of Nell."[22] Izaak Walton (d. 1683), author of *The Compleat Angler*, added a codicil to his will for the distribution of mourning rings to be engraved "A Friend's Farewell IW obi," the value of the rings to be thirteen shillings and fourpence each (Hunter 10). Sir Ralph Verney notes in 1685 that "Sir Richard Pigott was buried very handsomely . . . we that bore up the pall had rings, scarfs, hatbands, shame gloves of the best fashion." The rest of the gentry attending Pigott's funeral had mourning rings, and the servants wore special gloves.[23] In her will, novelist and playwright Mary Davys (d. 1732) left her impoverished sister clothing and an inexpensive mourning ring (totaling less than 5 pounds).[24] Of course, mourning wear could be very expensive, but it could all be reused and worn (as well as bequeathed). In his October 27, 1661, entry, Pepys humorously writes, "To church, my wife with me, whose mourning is now grown so old that I am ashamed to go to church with her." For his own mourning jewels, in 1703 Pepys ordered forty-six mourning rings at twenty shillings, sixty- two rings at fifteen shillings, and twenty rings at ten shillings. These rings were given to friends, family, servants, naval personnel, university colleagues, and others at the funeral, as was the standard of the time (Hunter 10). Those not present at the funeral had their rings delivered to them. By designating the more expensive rings for those closest to him (three classes of rings), Pepys reflects the practice of how family and friends could present themselves in society during the stages of mourning, as well as class distinctions.[25] Underscoring the need for class distinction, in his twenty-two page will (excluding numerous codicils), Horace Walpole, the 4th Earl of Orford, bequests a sizable "two hundred pounds for a ring" to each of his executors (totaling about £25,000 today).[26] With growing commodification and extravagance of mourning also comes the rise of sentiment in the period, reflected by Calvinist writer James Hervey's *Meditations Among the Tombs* (1746), which describes a brief epitaph as "*Concise* enough, to be the Motto for a mourning Ring."[27] Rings were memorialized in personal, intimate ways, such as Samuel Johnson's response to the death of his wife Tetty (d. 1752): "Johnson removed her wedding ring, preserving it, for the rest of his life, in a little wooden box in which he noted the dates of her marriage and death."[28] Johnson transforms the wedding ring into a mourning ring, engraving and encoffining it in a drawer next to his bed, where he can have recurrent access to the memorial.

These practices necessarily seep into early fiction. While Donne's metaphysical poetry incorporated burial and relic practice in the Renaissance, the emerging novel provides a narrative strategy for the tangibility of the relic: it can be encased and hidden, disappear, reappear, be caressed and kissed. As an exemplar, *Clarissa* demonstrates what we can learn about these rituals from detailed fictional wills and inheritance plots. While the novel houses the frequency of these cultural practices and emotional economies, many of the relics are only briefly mentioned; they act as floating signifiers of sentimental and/or consumer norms. A brief mention of mourning rings in *Moll Flanders* inspires a more detailed vignette in its adaptation. When Daniel Defoe's Moll Flanders steals a bundle from a burning house, she opens it to find numerous pieces of jewelry, including "a little Box of burying Rings."[29] In T. Read's period abridgment/adaptation of *Moll Flanders* titled *The Life and Actions of Moll Flanders* (1723), he depicts Moll's death and burial (and gives her the common name "Elizabeth Atkins"). He also includes her last will and testament, as well as the description of her respectable funeral, which mentions the funeral attendance of "above one hundred and twenty other Persons, who had gold Rings given them, with these Words engrav'd in them, *Memento mori. Elizabetha Atkins obit 1722.*"[30] When Mrs. Grizzle prepares for her death in Tobias Smollett's *The Adventures of Peregrine Pickle* (1751), we learn that "She bequeathed to commodore Trunnion one thousand pounds, to purchase a mourning ring, which she hoped he would wear as a pledge of her friendship and affection."[31] This expensive ring would most likely include diamonds and other precious stones.

As Frank notes, mourning rings, brooches, and lockets convey "how much our understanding of death and our relationship to those who have died have changed over time" (119). With this, it is important to recognize that these intense rituals of sentimentality around mourning jewelry focused on both men and women. Colonel Morden's request to make a locket from Clarissa's hair "to wear next his heart in memory of his beloved Cousin" was not uncommon in the mid-eighteenth century. Ariane Fennetaux acknowledges these rituals "were carried by men who were just as keen as women to keep and cherish jewelry in remembrance of their loved ones."[32] Jewelers' lists of sales include numerous orders from men acquiring memorial artifacts. Hunter and Fennetaux both cite the archives of eighteenth-century Aberdeen jeweler John Ewen (or Ewan), which give numerous examples of careful orders for mourning jewelry placed by men (correspondence often referring to the enclosure of locks of hair from a departed loved one).[33] A 1780 commission sent by a Mr. Forsyth precisely describes a mourning ring and its design, noting that it is for a "gentleman to be done with an oval crystal with an urn of the hair inclosed. The urn may have a weeping willow tree hanging over it . . . on the side of the urn may be put the letters JMcL . . . Round the oval crystal there may be an edge in enamel to give it the effect of a woman sitting suitable to such rings. Around the shank of the ring must be put in enamel Janet MacLeane died 21 July 1780."[34] As we saw with Pepys's requests and others, mourning jewelry—for both men and women—had a largely social function of designating status, reaffirming social rank and hierarchy. Lou Taylor notes, "After the death of Charles I a great number of memorial rings worn by royalists, some of them with the King's head, a skull and the words '*Prepared Be to Follow Me.*' This set a fashion and thereafter mourning rings were used increasingly."[35] By the end of the century we see an embracing of sentiment, "creating communities of sympathy constructed around kindred feeling rather than kindred rank" (Fennetaux 33). Similar to mourning portraits and other forms of memorials, mourning jewelry's *vanitas* iconography changed to reflect this more sentimental practice.[36] And later, the popularity of the practice changes in the Victorian period. For example, the custom of giving out mourning rings at the funeral was not typical in the nineteenth century but was relegated to close family.[37]

While the period saw a renewed interest in the giving of mourning rings, I'm particularly interested in how these artifacts emerge as secular relics in the eighteenth century, often enclosing hair (whether of the deceased or bought by the jeweler from an unrelated person to use for hairwork). As hair does not decompose or grow after death, it has always been a tangible souvenir of a deceased friend or relative, a European practice in Elizabethan times (as Donne's "The Relique" signifies; the Church of England cleric was raised Catholic)[38] and renewed in popularity during the late seventeenth century. As the line "*A bracelet of bright haire about the bone*" suggests, Donne's lovers are conjoined in the grave, the speaker's forearm tethered with the woman's "bright haire." Used in both love and mourning tokens, hair was a powerful talisman. Yet the insignificance and superficiality of hair also speaks to its liminal meaning. Silver aptly asks, "As for the thing itself, that dalliable excrement, what could be more abject than hair? Life in death and death in life, unfeeling and continually expelled: clearly what matters are the attachments it affords; hair is a matter of how it modifies relationships. In life or death, objects like hair rings worn as the counterweights to the strands of power and affection stretched between actors. Such jewels are 'exhibited secrets,' private things that signify only because they are worn publicly" (250). For Silver, hair rings "exist at a specific junction of modernity, where early forms of consumer culture cross with traditional forms of love and death" (250). This emphasis on consumer culture is also recognized by Taylor, who remarks, "Amongst the fashionable and elegant [hair] was a more socially acceptable memento than bones or teeth."[39] Jewelers' trade cards from the period reflect the popularity of hairwork in mourning accessories. Norris Junior's card (c.1780) identifies him as "Goldsmith & Toyman" but also as a hairwork and mourning jeweler: "Devices in hair execut'd in the neatest manner" and "Mourning Rings expeditiously made" (fig. 4.1). The billhead (c. 1794) of Benjamin Laver, "Working Jeweller & Goldsmith, To his Royal Highness the Duke of Clarence," advertises "Mourning Rings & Devices in Hair, neatly Executed."[40] This slippage between abject commodity and cherished relic, between fashionable display and emotional symbol, speaks to the broader cultural work of changing mourning rituals in the period that this chapter interrogates.

It should be no surprise that hairwork sees a renewed popularity in the period. In *A Cultural History of Hair in the Age of Enlightenment*, Margaret K. Powell and Joseph Roach acknowledge that there is "an increasingly robust preoccupation with hair over the course of the long eighteenth century" and argue for "the central place of hair in Enlightenment cultural representations."[41] Wigs worn by both men and women were markers of class, gender, occupation,

Figure 4.1. Trade card, Norris Junior, Jeweller, c. 1780.
© The Trustees of the British Museum.

and religion. Whether false or natural (the popularity of wigs waxes and wanes in the period), hair spawned a tonsorial industry (powders, tonics, the need for hairdressers and the like). Hair in the period is "both a material fact of nature and social magic" (Powell and Roach) and likewise is fitting for both fashionable and sentimental mourning practices. Hair transcends the corruptible body and is a tangible, physical connection experienced by the living with the dead. Less perishable than flesh, hair from a corpse could be artfully manipulated and treasured, symbolizing both spirit and matter, memory and materiality. Jayne Lewis notes the superficial yet lively aspects of hair that assist "hair's apparent transformation in the Enlightenment from a saintly relic fraught with miracle to a tangible personal artifact tied to both the inevitable loss of a body and to the sentimental possibility of that body's figurative persistence in memory."[42] Hirsutal attachments can be intentional: they importantly bind mourner with the mourned, but also with observers who serve to acknowledge another's loss. Helen Sheumaker, in *Love Entwined: The Curious History of Hairwork in America*, recognizes that hairwork helped to validate the experience of mourning by providing a physical reminder that the mourner's pain was real and demonstrable to others.[43] Thereby, crinitory excrements receive cultural weight in the period. Christiane Holm's reading suggests that mourning locks achieve sentimental meaning only through the liminal, excretal qualities of hair (existing between matter and meaning): "The cut edge of hair in the material medium of remembrance marks the act of remembrance as the very moment when its natural status was transformed into a cultural status."[44]

Though Fennetaux remarks that placing hair in mourning rings decreased in popularity, we still witness the practice in the early nineteenth century. For example, in 1818, James T. Power, fearing he would die on a trip to Sierra Leone, wrote to his love Julia Woodforde describing his plans for a mourning ring to be made for her, entwining each of their locks of hair: "I purchased yesterday a Diamond Mourning ring, I will place a brade [*sic*] of your hair and mine in it. ON the inside I will inscribe James T Power died . . . leaving a vacancy for the date if this should be my fate shortly you will receive an account of the time and get it filled up and I have no doubt you will regard the ring with affection and wear it on my account."[45] These examples and others also indicate that sentimental practices that developed around mourning jewelry were not gender specific. Interlocked hair from two lovers generates notions of bodily connection, and intersects both love and grief. Lockets, rings, miniatures, and other material artifacts housing hair encapsulate the growing emotional economy centered on the body, loss, and mourning.

We have seen that jewelry acts as both love token and mourning memento, especially cherished if created with hairwork. In Elizabeth Justice's *Amelia, or, The Distress'd wife: a history founded on real circumstances. By a private gentlewoman* (1751), published on the heels of *Clarissa*, Justice describes the deathbed preparations of Amelia's mistress Mrs. Sweet, who has contracted smallpox. Her preparations include hair jewelry: "Two or three days before she died, she said to *Amelia*, My dear, you will oblige me to go to Mr. *Blunt's*, the Goldsmith, and desire him to make a Mourning Ring with a Christal, in a Lozinge; for I should love to see the Rings that I design for my friends."[46] Though not described, the lock of hair, we can assume, would be encased. In Helen Craik's anti-Jacobin *Adelaide de Narbonne, with memoirs of Charlotte de Cordet* (1800), Victorine opens a casket containing her dead parents' valuables, including their miniatures, but it is their hair jewelry that is most prized and intimate: she discovers "lockets containing the united ciphers of her parents, done in their own hair."[47] Along with rings and brooches, lockets would be common relics for the middle and upper classes. This mourning locket is a beautiful example of mourning hairwork such as Craik describes, including the blond hair of the "Countess Dowager of Home died 15 Jany 1784" (inscribed on the back) with her initials "ECJ" under a coronet of gold and pearls under glass (fig. 4.2). In Elizabeth Griffith's epistolary novel *The Delicate Distress* (1769), Lady Woodville describes a miniature portrait worn as a bracelet: "Charming lady Lawson! What an engaging countenance, what a quick sensibility in her looks, what an irresistible smile! I am not under a necessity of looking at my bracelet, to remind me that this portrait resembles lady Straffon: but lady Lawson is taller, thinner, and more of the brunette" (12). Later Lady Straffon gifts her portrait for a wedding gift: "I had the pleasure of placing my picture in a bracelet, upon lady Somerville's arm" (120). Though hair is not an element, these rare descriptions of bracelet miniatures in a novel represent this fashionable practice. Yet it is the intertwining of hair and miniature that creates unique eighteenth-century secular relics (and because of their delicate nature, few have survived). Watch fobs, hair chains, and bracelets were common.[48] Hallam and Hockey note that "hair could be worked into tubes, ropes and bands intended to encircle the wrist," and that these "play upon both the visual and tactile aspects of the memory 'object'" (139). The hair bracelet with portrait of Constantine Phipps (c. 1770) would be typical for the period: worn by women, bracelets often included a male miniature portrait and were woven with the recipient's hair (fig. 4.3).

As hair becomes a necessary element of miniatures and jewelry, we also see the aesthetic change later in the period to representing the grieving versus the

Figure 4.2. Mourning locket, Countess Dowager of Home, c. 1784. Engraved and enameled gold set with seed pearls and hair. England. © Victoria and Albert Museum, London.

symbols of death. For both mourning miniatures and jewelry, Neoclassical forms, Christian iconography, and archetypal images of mourner, urn, tomb, and willows would be chosen from sample devices by the maker and client, then individually commissioned and designed.[49] This was quite a shift in symbology, as Lou Taylor recognizes: "In medieval Europe under the influence of the Catholic church, when representations of the terrors of death and hell were dwelt upon in horrific detail, death's head and skeleton motifs were frequently used in memento mori jewellery, gradually going out of use from the eighteenth century" (188). While *vanitas* (of Dutch origin) skulls or skeletons are still evident through

Figure 4.3. Woven hair bracelet with portrait on ivory of Constantine Phipps, c. 1770. John Smart, jeweler. Courtesy of the Edward B. Greene Collection, Cleveland Museum of Art (1951.437).

midcentury, memorial jewelry begins to reflect more sentimental imagery. Holm observes that in eighteenth-century iconography, "the urn . . . is just an euphemism for the traditional skeleton" (142), a reinforcement of its bony connotations and artistic medium. Frank notes this change in symbology from skeletons and crossbones to lamenting grievers, urns, and willows: "The shift from an emphasis on death itself to one on the lamentation for the dead developed in tandem with the new enthusiasm for family life being celebrated by portrait miniatures" (121). This development of the affectionate family dynamic resulted in a greater expression of grief over loss. Frank explains, "Mourning scenes, often limned on the reverse of portraits, keep the absent family member ever present" (121). The majority of eighteenth-century mourning lockets, brooches, and rings include this iconography, such as this typical example of a ring with weeping mourner, willow tree, and urn (fig. 4.4). Hair is actually a part of the portrait's medium on enamel, mixed with paint and chopped for design work. Though somewhat less common, "male mourners also appeared in allegorical scenes as stand-ins for the bereaved" (Frank 124). Middleton's trade card with its female mourner, willow, and urn exemplifies the griever's iconography as well as the growing control undertakers had over furnishing funerals with "mourning rings & other Jewellery made to order on the shortest notice" (fig. 4.5).

Figure 4.4. Mourning ring, c. 1784. Engraved gold with hair. England, late eighteenth century. © Courtesy of the Victoria and Albert Museum, London.

Underscoring the prevalence of the imagery, graveyard iconography is notably depicted in poetry and meditations, but the novel as well, such as Radcliffe's description of a mourning La Luc in *The Romance of the Forest* (1791): "[M. Verneuil] would have addressed La Luc, when he perceived him at a distance leaning against a rustic urn, over which he drooped, in beautiful luxuriance, the weeping willow."[50] M. Verneuil approaches the mourning urn and sees it is inscribed, "To the memory of Clara La Luc, this urn is erected on the spot which

Figure 4.5. Trade card of Richard Middleton, coffin maker and undertaker, c. 1760–1818. © The Trustees of the British Museum.

she loved, in testimony of the affection of a husband." This iconography, which will diminish in the nineteenth century, is most commonly found in jewelry from the period, as is often noted in fiction. In Charlotte Palmer's epistolary novel *It is, and it is not a novel* (1792), the sentimental heroine Miss Heartfield (named appropriately), faints when she meets Miss Digby, who proceeds to remove Miss Heartfield's gloves for "greater relief," noticing with surprise, "On her first finger was a mourning ring, with a weeping willow, drooping over an urn!—I was afraid my features unfortunately bore some resemblance to a friend she has lost."[51] Later when she finally receives a letter from her lover who has not corresponded in weeks, Miss Heartfield chooses to go "into the garden to read it, seating herself on a bench under a willow, which she had frequented in the absence of her perfidious lover!" (316). In deep mourning and seated under a willow, she is the exact portrait of sentimental memorialization.

Visiting the grave, keeping miniature reliquaries of the deceased, and encasing locks of hair represented the need for a tangible location of the dead. Coupled with the cult of sensibility and a more modern notion of a bonded family, intense mourning and its material representations increasingly became part of eighteenth-century culture. Frank explains that "The emphasis on weeping over loss reflects enlightenment ideas about humankind's innate moral compassion. By the late eighteenth century, sentiment—the expression of feelings with open outbursts of emotion, especially tears—was seen as a desirable trait. The exaggerated rhetoric of the sentimental had a profound impact on mourning iconography" (121–23). A powerful example is Johann Wolfgang von Goethe's sentimental novel *The Sorrows of Young Werther* (1774), which directly influenced the design of memorial jewelry.[52] The novel's etchings by J. R. Smith include urns, weeping trees, and grave visitations. Numerous iterations of Charlotte grieving over the tomb of Werther circulated in the late eighteenth century, including mourning jewelry, such as an English locket (c. 1775) with painted ivory, chopped hair, and plaited hair surrounding the portrait (fig. 4.6). Though the hair in mourning jewelry was not necessarily made from that of the deceased,[53] the amount of hair used for the plaiting surround of *Charlotte at the Tomb of Werther* could suggest that the recipient indeed commissioned the locket and included the deceased's hair.

As we have seen, the passionate need to keep the body of the beloved near is symbolically expressed by hair, the residue of the living. Locks of hair were a common British memento in the eighteenth century, given among relatives, friends, and lovers and typically encased in brooches, rings, and lockets, a practice particularly reflected in fiction. In Fielding's *Amelia*, Captain Booth describes how he has cherished Amelia's lock of hair: "The most valuable of all to me was a lock of her dear hair, which I have from that time worn in my bosom" (101–2). In Georgiana, Duchess of Devonshire's *The Sylph* (1779), the unhappily married heroine Julia gifts the sylph, her secret moral guide, a sentimental hair locket. She writes her sister, describing its design and aesthetic with intricate detail:

> This morning I dispatched . . . the most elegant locket with hair that you ever saw. May I be permitted to say thus much when the design was all my own? Yet, why not give myself praise when I can? The locket is in the form and size of that bracelet I sent you; the device, an altar on which is inscribed these words: *To Gratitude*, an elegant figure of a woman making an offering on her knees, and a winged cherub bearing the incense to heaven. A narrow plait of hair, about the breadth of penny ribbon, is fastened on each side of the locket near the top of three diamonds and united with a bow of diamonds, by which it may hang to a ribbon. I assure you, it is exceedingly pretty. I hope the Sylph will approve of it. I forgot to tell you, as the hair was taken from my head by

Figure 4.6. Mourning locket, c. 1775. *Charlotte at the Tomb of Werther*, after an etching by J. R. Smith, inscribed *May Saints Embrace Thee with a Love Like Mine*. © Courtesy of the Victoria and Albert Museum, London.

> your dear hand before I married, I took the fancy of putting the initial J.G. instead of J.S. It was a whim that seized me, because the hair never did belong to J.S.[54]

This intimate gift symbolizes Julia's intense feelings for and gratitude toward the Sylph. Realizing its associations with love, Julia attempts to distance this intimacy by having the locket engraved with her maiden name initials. The kneeling female figure, altar, and incense reflect sentimental iconography, while the cherub connotes a Rococo figure common on eighteenth-century mortuary sculpture. Exceptional is the detail the duchess provides, demonstrating the significance of these intimate gifts of hair jewelry, especially among the aristocratic classes.

Jane Austen's *Sense and Sensibility* (1797; pub. 1811) includes the famous scene of Marianne Dashwood cutting a lock of her hair as a gesture of love and gifting it to Willoughby, but the description of Edward Ferrars' "promise" ring closely aligns it with a mourning piece. When Marianne notes the new ring on Edward's finger, a ring "with a plait of hair in the centre," he blushes and announces that the hair is his sister's, despite Marianne noticing that it seemed darker than his sister's: "'I never saw you wear a ring before, Edward,' she cried. 'Is that Fanny's hair? I remember her promising to give you some. But I should have thought her hair had been darker.'"[55] Realizing how intimate the giving of hair is, he is forced to lie: "The setting always casts a different shade on it you know" (74). Both Dashwood sisters assume the lock is Elinor's. Marianne thinks the hair has been freely given by Elinor, while Elinor "was conscious must have been procured by some theft or contrivance unknown to herself" (an erotic gesture invoking Pope's *Rape of the Lock*).[56] Eventually, they discover the ring and lock of hair were given by Lucy Steele (who also carries a miniature of Edward). In fact, it is a major clue that Edward and Lucy are secretly engaged. Unlike mourning hair jewelry, tokens of love were usually not engraved with a name, and so could be anonymous.[57] Later, Lucy tells Elinor, "If he had but my picture, he says he should be easy. I gave him a lock of my hair set in a ring when he was at Longstaple last, and that was some comfort to him, he said, but not equal to a picture" (102).[58] Austen's novel demonstrates that by the end of the century hair was a part of sentimental gift giving and aesthetics, not just relegated to funerary purposes. Edward's wearing of the hair ring also signifies that men were deeply involved in this sentimental practice and publicly displayed tokens of affection.

In Charlotte Smith's *Celestina* (1791), a locket entwines the hair of three loved ones and is cherished by the novel's heroine, Celestina, an orphan "adopted" by the Willoughby family: "[Celestina] took out of her dressing box a locket, in which [Willoughby's] hair was interwoven with that of his mother and sister, and which she had been used as a child to wear around her neck. She looked at it a moment, and remembered a thousand circumstances that brought the tears again into her eyes. She kissed it; she put it to her heart; and that soft heart melting at the tender images this slight memorial presented to it, the resentment which her pride had made her feel the evening before was forgotten."[59] It is striking that the hair locket is described as a "memorial" though it contains the hair from three living relatives. Similar to how the miniatures previously discussed were treated, it is caressed, kissed, and pressed to the heart as if it has pious significance. Smith details the power of this material representation of the body and person, and with this we can better understand the significance human remains will have in mourning practices. The sight of hair inspires emotion, even tears, appropriate responses for a person of sentiment and sensibility.

The giving of rings is so common that by the end of the century it becomes cruelly symbolic, such as the scene in Elizabeth Inchbald's sentimental romance *A Simple Story* (1791).[60] Entering what is seen as an unnatural marriage to a much older father figure, her guardian Lord Elmwood (a former Catholic priest named Dorriforth), Miss Milner receives an unexpected gift from her husband during their wedding ceremony: "On that first wedding-day, that joyful day, which restored her lost lover to her hope again; even on that *very* day, after the sacred ceremony was over, Miss Milner—(with all the fears, the tremors, the superstition of her sex)—felt an excruciating shock; when, looking on the ring Lord Elmwood had put on her finger, in haste, when he married her, she perceived it was a—MOURNING RING."[61] The rite of marriage thereby turns into an implied funeral rite when he mistakenly gives her a mourning ring. It is as if at this moment in the plot Elmwood transforms into a tyrant, and the mourning ring becomes an omen of their future unhappiness. With this symbol, the novel foreshadows Miss Milner's (now Lady Elmwood's) eventual fall (resulting from her affair), banishment, and agonizing death. Inchbald symbolically associates her with death throughout the novel (such as an illness that causes a "palsy over her frame" and a "paleness in her face"), yet it is the mourning ring that most powerfully speaks to Miss Milner's tragic fall and death.

Modern Western cultures often used hair as a memento of the dead, but it becomes a material phenomenon in the late seventeenth and early eighteenth centuries. As mourning jewelry weaved together fashion, display, and sentiment, hair became an integral artistic medium, chopped, dissolved into paint, formed into initials and writing, and decoratively plaited. Frank effectively explains the symbology of hair jewelry: "That passionate need to break through the barrier between the living and the dead and keep the body of the beloved near is symbolically expressed" by the "lock of . . . plaited hair—a residue of . . . identity."[62] Hair jewelry materially preserves identity and transcends the mortal condition. Miniatures and jewelry that encase locks of hair bind the living to the dead. Receivers can easily wear, touch, and gaze on a seemingly immortal body part. Pointon notes that while on one level hair jewelry is a "mummification" of desire (a form of sexual exploitation and containment of hair), it is also "the materialization of memory as personalized, and miniaturized."[63] "Participating in the collection of precious objects that is a requisite part of the *vanitas* tradition, as also of the museum . . . hair jewelry materializes grief as secular reliquary and micro-museum," Pointon recognizes.[64] Similar to wax objects and anatomy specimens, hair memorials can also function as commodified and circulated human remains. Period instruction manuals, directed to mainly women who practiced hairwork, organize hair jewelry into types, are typically numbered (like natural history specimens), and "invite speculation upon two absent bodies, that of the donor of the substance and that of the wearer."[65] Yet they also function as immortalized tokens, manipulated and designed. Shown in this chapter, many of the plaited and twisted forms are circular (bracelets, rings, brooches) and suggest a type of eternity. With its intertwined coils, plaited and twisted, hairwork materially demonstrates the repetitious remembrance of mourning. An example of this is Joseph Severn's miniature of John Keats (c. 1818) (fig. 4.7). Though completed when the poet was still living, the portrait functions as a memorial. Housing an inset that encases Keats's coiled hair, the miniature serves as a reliquary, preserving a precious remnant of the poet's body.[66] The display and circulation of Keats's hair replays the well-known conceit in his poem "Lines on Seeing a Lock of Milton's Hair" (1818): "A lock of thy bright hair—/ Sudden it came, / And I was startled when I caught thy name / Coupled so unaware" (3.8). Milton's posthumous circulation is reported by Physician Philip Neve in *Narrative of the Disinterment of Milton's Coffin* (1790). Neve relates that the discovery of Milton's grave at St. Giles led to church officials opening the coffin to see the corpse (a common practice detailed in Chapter 2 of this book). Seeking relics of the great poet, attendees pulled and knocked out Milton's teeth, took a piece of his jaw, and as there was "a great quantity of hair," clumps were pulled from the skull. Neve explains that Mr. Ellis, "an ingenious worker-in-hair" thought that "it would be of great advantage to him to possess a quantity of Milton's hair."[67] Bits of the coffin and other body parts (such as a leg bone, ribs, and even one hand) were also removed, and later, admission would be charged to view the corpse, but it was the hair that would be famously commodified and fetishized. Pieces of shroud and hair were purloined by eager tomb hunters. Lorna Clymer describes that his remains were spread among private collections, "religiously preserved [as] precious relic[s]."[68] As he explains to friend Benjamin Bailey, Keats had lately come across "a real authenticated Lock of *Milton's Hair*" at the home of his friend Leigh Hunt, and was thereby inspired to write the poem.[69] (Similar to Walpole, Hunt would go on to collect the locks of Swift, Johnson, Shelley, and numerous other literary dignitaries).[70]

Figure 4.7. Joseph Severn, *John Keats*, c. 1818. Portrait enclosing hair. Watercolor on ivory. 108 × 83 mm. © Fitzwilliam Museum, University of Cambridge.

Accompanying portraits, encased in rings and lockets, and used in art, locks of hair (and chopped hair) formed various memorials. In Charlotte Smith's *Ethelinde, or The Recluse of the Lake* (1789), a button memorial (c. 1740) found on the sleeve of the injured soldier Montgomery (during the Seven Years' War) is described by his wife: "It was a button set with brilliants, that had once belonged to Lord Pevensey; and which, as the diamonds surrounded a cypher formed of her hair, had been, after his Lordship's death, given by my mother to Montgomery."[71] (The rules for first mourning banned shiny or jeweled buttons, so this may have been a memento for half mourning.) As Smith's valuable button signifies, common memorials for the upper class often included hair and jewels. Though the relationship of mourning jewelry to masculinity is complex, for the majority of the period a hair brooch, ring, or locket depicting a weeping woman or encasing a lock of hair could be worn by a man without undermining his masculinity.[72] The period also sees bejeweled mourning swords, watches, toothpick cases with hair insets, snuffboxes with hair insets, shoe and knee buckles with hair insets, all worn or carried by men. Another Laver trade card reflects this practice, advertising that he is a "Fancy Worker in Diamonds," including for "Swords, Buckles." In *Luxury and Pleasure in Eighteenth-Century Britain*, Maxine Berg begins her book with a description of natural philosopher and theologian Joseph Priestley's household items, destroyed in a 1791 mob raid. Among these are "New forms of ornamental ware and medallions," including "mourning buckles."[73] Berg goes on to say, "This inventory provides an arresting image of the material culture of an established middling-class intellectual at the end of the century" (3). By the 1780s, the atmosphere of alarm about luxury and increased consumption seems to have diminished somewhat. And the inclusion of valuable mourning buckles in the Priestley inventory does reflect common mourning objects for men. Despite their diminutive size, these buckles could also be treated as personalized relics, engraved with death dates and encasing hair. For example, a gold mourning buckle (c. 1725), most likely for the knee, engraved with gold thread/wire lettering and hair commemorates the death of a three-year-old girl, the wire inscription reading "Hannah Kill/Dyed 17 AP Aged 3 Years & 8 Days 1725" (fig. 4.8). Surprisingly, even an enameled toothpick case (c .1780) can be transformed into a sentimentalized object. A distinctive case commemorating the death of Robert Henley Ongley, 1st Baron Ongley (c. 1720–1785), who sat as the Whig member of Parliament for Bedford, is set with diamonds and plaited hair under glass (fig. 4.9).[74] This particular example is enameled, but toothpick cases were often made of ivory, bone being an element that I discuss next in Chapter 5. The practice of gifting toothpick cases may have started with Charles II, who left a mourning ring and "tooth-pickcase" for his father.[75]

These varieties of mementos attest to the fact that in the eighteenth century, mourning jewelry and accessories became less about *memento mori* and *vanitas* motifs and more about possessing an actual relic or symbolic representation of the dead with which to attach loss and grief, reflecting the social and emotional rules of mourning. In particular, hairwork in mourning jewelry and miniatures was an expensive, sentimental, bodily reminder of the deceased that would cul-

Figure 4.8. Mourning buckle, c. 1725. Set with woven hair underneath panels of rock crystal, the gold wire inscription reads "Hannah Kill / Dyed 17 AP Aged 3 Years & 8 Days 1725." © Victoria and Albert Museum, London.

minate in the hair industry of the nineteenth century. Representing the presence of the dead, hair is a permanent material that can either be braided or left simply as an encased, preserved lock. And like bone, hair can be chopped, glued, and manipulated into artificial forms that deflect from its reference to the dead body. Worn on the warm, *living* body, hair jewelry and other relics can be caressed and gazed at; they are artifacts of secular reliquary and remembrance. With growing

Figure 4.9. Toothpick case, c. 1780. In memory of Robert Henley Ongley, 1st Baron Ongley. Enameled, champlevé, engraved gold, set with diamonds, mirror, and plaited hair under glass. London. Gilbert Collection (loan). © Victoria and Albert Museum, London.

unease at the process of physical decay, the culture desired more aesthetic forms of preservation and grieving, and the novel catalogs this shift, uniquely contributing to a cultural history of materialism. Whether as mementos or luxury items, hair jewelry necessarily finds its way into narratives, beautifully encased and immortalized.

CHAPTER 5

Osseous Matter

Bones, Relics, and Mourning Miniatures

> I shall become like the mortals on whose reliques I now gaze, and, like them too, I may be the subject of meditation to a succeeding generation.
>
> —Ann Radcliffe, *The Romance of the Forest*

As Clarissa meticulously prepares for her death following her rape, Lovelace's friend Belford writes to him that, "The lady has been giving orders, with great presence of mind, about her body! directing her nurse and the maid of the house to put her in the coffin as soon as she is cold. Mr. Belford, she said, would know the rest by her will" (8:12, letter 2). Though devout, Clarissa is anxious about her posthumous state and expresses concerns about the treatment of her corpse ("giving orders . . . about her body!"). I have been arguing throughout this book that this focus on the corporeal reflects shifting mourning practices in the period. In particular, Clarissa's cherishing of a portrait miniature as she prepares for her own death signifies the sentimental attitudes I will be discussing in this chapter. Belford continues: "She has just now given from her bosom, where she always wore it, a miniature picture, set in gold, of Miss Howe. She gave it to Mrs. Lovick, desiring her to fold it up in white paper, and direct it, To Charles Hickman, Esq. and to give it to me, when she was departed, for that gentleman. She looked upon the picture, before she gave it her—Sweet and ever-amiable friend!—Companion!—Sister!—Lover! said she—and kissed it four several times, once at each tender appellation" (8:12, letter 2). Ensuring that Anna Howe will have her miniature returned, Clarissa begins bequeathing personal, sentimental items before her death. She takes the miniature from her warm "bosom," and caresses, kisses, and speaks to the portrait, a performance with sentimental objects that we have seen repeatedly enacted in fiction. Later in her will, Clarissa underscores the miniature's significance as she bequeaths it to Anna's fiancé: "I bequeath to the worthy Charles Hickman, Esq. the locket, with the miniature picture of the lady he best loves, which I have constantly worn, and shall continue to wear next my heart till the approach of my last hour. It must be the most acceptable present that can be made him, next to the hand of the dear original" (8:105, letter 27). Next, she bequeaths her own portrait as a mourning gift to her cousin Colonel Morden: "I desire him likewise to accept of the little miniature picture set in gold, which his worthy father made me sit for to the famous Italian master whom he brought over with him; and which he presented to me, that I might bestow it, as he was pleased to say, upon the man whom I should be one day most inclined to favour" (8:105, letter 27). These moments dramatize the important role of miniatures in sentimental practices. They function as visual and physical substitutes for identities and bodies and are transformed into meaningful mourning objects. Clarissa's concern for how her dead body is treated is intimately tied to these visual representations: standing in for the decaying flesh, tangible portraits possess an aesthetic immortality and are proper ciphers for mourning.

Earlier, we noted Lovelace's alignment of Clarissa's skin with wax. In another letter to Belford, Lovelace describes Clarissa's body in a Pygmalion fashion, as if it is an ivory statue: "This was mine, my plot! and this was all I made of it!—I love her more than ever!—And well I may!—never saw I polished ivory so beautiful as her arms and shoulders; never touched I velvet so soft as her skin: her virgin bosom—O Belford, she is all perfection! then such an elegance!" (4:375, letter 59). Lovelace also invokes the traditional comparison of ivory and teeth: "If you love to see features that glow, though the heart is frozen, and never yet was thawed; if you love fine sense, and adages flowing through teeth of ivory and lips of coral" (3:329, letter 63). Fetishizing her skin, and rhetorically reversing the Galatea myth (Pygmalion's statue come alive), Lovelace reinforces the disembodiment and rhetorical dissection of Clarissa: she is reduced to "polished ivory," "arms," "shoulders," "skin," "bosom," "lips," and "teeth." Historically, ivory had often been used as a metaphor for pale, flawless skin in art and literature. Lovelace's ivory comparison perhaps is intentional, referencing Shakespeare's *The Rape of Lucrece* ("her breasts, like ivory globes circled with blue").[1]

This intersection of sentimental mourning, cherished portraits, and fetishized body parts in *Clarissa* reflects the broader subject of this chapter that will illuminate the significance of mourning miniatures in the culture and interrogate the material often used for miniature portraits: bone. As with hair, bone was an important, permanent material for secular relics, regarded as a "powerful memory object" (Hallam and Hockey 134). My larger aim is to correlate the cultural circulation of osseous matter (in morbid collections, curiosity cabinets, and public exhibitions) with mourning practices, further uncovering how the culture's growing acceptance of secular relics is vividly translated and dramatized in fiction.

In Eliza Haywood's *The History of Miss Betsy Thoughtless* (1751), the ubiquity of miniatures is displayed when Trueworth's portrait is described by Thomas: "'Never was there a more perfect likeness;–he is drawn in miniature; I believe, by the size of the piece, intended to be worn at a lady's watch;—but I looked on it through my magnifier, and thought I saw his very self before me.'"[2] He is so impressed with the "excellence of this artist" that the narrator remarks Thomas desired that "his favourite mistress's picture should be drawn." As Haywood points to, even a miniature has the power to conjure the physical subject: "I saw his very self before me." In Henry Fielding's final novel *Amelia* (1751), a similar pictorial embodiment is described by Captain Booth, who laments the loss of the miniature: "Next to *Amelia* herself, there was nothing which I valued so much as this little Picture: for such a Resemblance did it bear of the Original, that *Hogarth* himself did never, I believe, draw a stronger Likeness."[3] It is the miniature's unique form and size that allow it to do justice to the individual and conjure (and encase) the body. Jakub Lipski identifies in *Painting the Novel: Pictorial Discourse in Eighteenth-Century English Fiction*, "The miniature, in contrast to large-scale aristocratic portraits, functions as an index of individual identity. It can be treasured as a keepsake of the beloved one, rather than as the monumental testimony to abstract nobility."[4] For early modern eyes, the possession of miniature portraits, as Orest Ranum argues, expressed a "desire for intimacy, for bodily contact, no less significant than the desire to be buried along with the beloved."[5] Yet the miniature also slips into object territory: in the eighteenth century, miniatures were displayed in curiosity cabinets along with petrified reptiles, carved ivory, fossils, "pickled monsters," and crucifixes (Stafford 242). This defiance of the miniature to be simply relegated as mourning object represents the fluidity and boundary crossing of most secular material forms in this period. Hallam and Hockey aptly describe this shift: "Through this process of boundary crossing, the memento mori could be endlessly shuffled so that its functions once residing within the sphere of memory could be over-ridden and replaced with other social and cultural meanings" (64). This functional slippage between public and private, religious and secular, displayed and secreted, speaks to the mourning miniature's fetishized purpose in fiction.

Treasuring these intimate keepsakes, both eighteenth-century men and women wore the miniature portrait of a loved one. In *Hanging the Head: Portraiture and Social Formation in Eighteenth-Century England*, Marcia Pointon observes that "wearing miniatures entered popular discourse as a topos of affective private engagement."[6] When gifted, the miniature stands in for the person, but also for the person's body. Pointon describes this important reification: "In secular and bourgeois society, portraits work actually and metaphorically to secure a connection between an absent person and the viewer" (57). Thus, Clarissa's bequeathing is deeply symbolic; the miniature symbolizes her absent body. Often miniatures were carried in a pocket or, as Haywood mentions, on a watch chain, sometimes made of braided human hair. The public display of the miniature on the body signaled not only a family's wealth and status, but affection. This performance and display has been linked by Graham Reynolds, Joe Bray, and Katherine Coombs to the rise of public exhibitions (such as the Royal Academy's) toward the end of the century.[7] Identifying this role of public performance, Holm remarks, "Mourning jewels are exhibited secrets" (140). They do indeed symbolize a condensed form of loss which only performs and signifies when connected to the body. This narrative of loss is intimate and personal while also exposed and observed. In her seminal study *On Longing: Narratives of the Miniature, the Gigantic, the Souvenir, the Collection*, Susan Stewart describes the significance of miniaturized worlds: "The miniature, linked to nostalgic versions of childhood and history, presents a diminutive and thereby manipulatable, version of experience, a version which is domesticated and protected from contamination. It marks the pure body, the inorganic body of the machine and its *repetition* of a death that is thereby not a death."[8] Stewart's linking of the miniature with temporality and death, as well as the body and touch, and her explanation of the forms of "preservation" the miniature can take, assist my reading of secular memorials and their significance in the period.

Mourning miniatures[9] haunt the eighteenth-century novel. Standing in for the dead body, they are caressed, kissed, and worn by the mourner who desires close proximity to a lost loved one. Samuel Richardson's *The History of Sir Charles Grandison* (1753) includes a sentimental scene that reflects the emotional engagement with a miniature. When Sir Charles's mother dies, he is given a silver casket containing her jewelry. As he and his sisters look through it, Charles discovers a valuable miniature:

> Then opening the purse, he found five coronation-medals of different princes in it, and several others of value; a gold snuff-box, in which, wrapt in cotton, were three diamond rings; one signified to be his grandfather's; the two others, an uncle's and brother's of Lady Grandison: but what was more valuable to him than all the rest, the ladies said, was a miniature picture of his mother, set in gold; an admirable likeness, they told me; and they would get their brother to let me see it. Neglecting all the rest, he eagerly took it out of the shagreen case; gazed at it in silence; kissed it; a tear falling from his eye. He then put it to his heart; withdrew for a few moments; and re-turned with a cheerful aspect.[10]

Miniatures are valued for the quality of the portrait—an "admirable likeness" assures the recipient that it has emotional value and deserves to be, as Charles performs, cherished, kissed, and placed near the heart. If not worn or secreted in a pocket, miniatures are encoffined, placed in caskets and resurrected at special moments for the owner's comfort. Frank remarks that "The growth of increasingly private, child-centered families made loss harder to bear and contributed to the miniature's popularity as a token of mourning . . . most commemoratives, whether posthumous portrait miniatures or tiny scenes of weeping mourners, were private tokens that emblematically kept the absent family member within the circle of the living" (7). And as we will see, in fiction mourning miniatures are not taken by artists from the corpse nor do they typically depict the subject as dead. The features of the *living* person are most desired along with a memento from the body (typically hair).

As memorials begin to focus on the mourner, numerous miniatures depict figures in mourning. *Maria, Duchess of Gloucester when Lady Waldegrave* (c. 1765) shows her wearing widow's weeds (and is a copy of her full-size portrait by Sir Joshua Reynolds) (fig. 5.1). She is in mourning for her first husband, James Waldegrave, who died in 1763. The portrait exemplifies a common melancholic pose of the period: a sorrowful female with body turned to the left, eyes not facing the viewer but instead raised to heaven. Though the portrait does not contain hair, bone, similar to its use for mourning rings, is an implicit element of this miniature: it is composed of watercolor on ivory, and Maria's mourning dress would be shaped with bone.

Not surprisingly, numerous mourning miniatures appear in fiction in the second half of the century, as the sentimental mode had taken hold of the culture. For example, in Elizabeth Griffith's highly sentimental novel *The Delicate Distress* (1769), the young heroine Charlotte Beaumont, whose birth has been

Figure 5.1. William Marshall Craig, *Maria, Duchess of Gloucester when Lady Waldegrave*, c. 1800–1827. Miniature. Watercolor on ivory. 6.4 × 5.3 cm. Royal Collection Trust / © Her Majesty Queen Elizabeth II 2021.

kept secret, learns of her true parentage. Typical of these plots, miniatures play a revelatory role, and they are often encased and enshrined by damsels in distress. On reading a cruel letter from her stepmother that purports to send her to a convent, Charlotte is in emotional distress and finds comfort in her parents' portraits:

> Full of these warm, and natural apprehensions, the half-distracted Charlotte flew to the cabinet, which, like Pandora's box, contained a thousand ills, and, with a trembling hand, unlocked it. The first objects that presented themselves to her view, were miniature portraits, of her father and mother.—She gazed, with joy and wonder. Never had she beheld such striking beauty, of both kinds; the manly, and the mild.

> She kissed, embraced, wept over them; nay, knelt to them, implored their pity and protection, and, in one moment, was inspired with more respect and tenderness, for those inanimate figures, than she had ever felt for her supposed mother; though gratitude and esteem had answered all the purposes of filial affection, in her gentle nature.[11]

Similar to Charles Grandison's, Charlotte's response will prove typical for fiction: the tangibility of miniatures, as they stand in for the absent and dead body, is stressed. Charlotte viscerally responds to the images: kissing, embracing, weeping, and kneeling. Though secular memorials, these objects slip into devotional territory as Charlotte reliquizes the mourning miniatures, kneeling before them and begging for "pity and protection." Along with the characters previously discussed, Charlotte participates in an intimate, tactile, secret ritual enabled only by the diminutive miniature. At the end of the novel, Charlotte is accepted by her father, but she dies before she can be released from the convent where she has been raised. As Charlotte had found with her parents' portraits, her grieving lover finds comfort in Charlotte's miniature, given to him as a gift from her brother: Lord Seymour describes that he "gaz[ed] intently, upon Charlotte's picture . . . I was so much softened, by the object then before me, my angel Charlotte's face! [I] burst [. . .] into tears" (222). The portrait now functions as a mourning object, endowed with sentimental qualities.[12]

In Ann Radcliffe's *A Sicilian Romance* (1790) we witness the power of the miniature to conjure the dead. While arranging some papers in the drawers of a cabinet, the heroine Julia is mesmerized by a mysterious portrait she discovers: "It was a miniature of a lady, whose countenance was touched with sorrow, and expressed an air of dignified resignation. The mournful sweetness of her eyes, raised towards Heaven with a look of supplication, and the melancholy languor that shaded her features, so deeply affected Julia, that her eyes were filled with involuntary tears. She sighed and wept, still gazing on the picture, which seemed to engage her by a kind of fascination. She almost fancied that the portrait breathed, and that the eyes were fixed on hers with a look of penetrating softness."[13] With "eyes, raised towards Heaven" and "melancholy languor," this description closely resembles the mourning portrait of the Duchess of Gloucester (see Figure 5.1). When Julia later shows the portrait to her governess, Madame de Menon, she learns "that she had wept over the resemblance of her mother," Louisa Bernini, her father's first wife (28). Radcliffe's personification of the portrait underscores the ability of the miniature to conjure life: it "engage[s]" Julia, "fixe[s] it eyes on hers with "penetrating softness," and seems to "breathe." Though Julia and Emilia think their mother is dead, and thus the miniature acts as a memorial, its personified description foreshadows the ending of the novel, when Louisa, who has been imprisoned, is found quite alive.

In fact, Radcliffe later replays the importance of the mourning miniature to the plot in *The Mysteries of Udolpho* (1794). In a voyeuristic scene, Emily St. Aubert describes her father's intimate engagement with a mysterious portrait: "The rays of light fell strongly upon [a miniature picture], and she perceived it to be that of a lady, but not of her mother. St. Aubert gazed earnestly and tenderly upon this portrait, putting it to his lips, and then to his heart, and sighed with a convulsive force. Emily could scarcely believe what she saw to be real. She never knew till now that he had a picture of any other lady than her mother, much less that he had one which he evidently valued so highly."[14] St. Aubert's experience signifies that miniatures have the ability to console the mourner, who can intimately touch and manipulate the material symbol of the body at whim (since it is portable). Similar to Sir Charles Grandison and Charlotte Beaumont, here St. Aubert treats the portrait as worshippers would a relic: it is kept hidden in a case, accessed at deeply personal moments, and caressed, kissed, and placed at the heart. (And though the novel's setting is Catholic France and Italy, St. Aubert and Emily do not distinctly practice Catholicism.) While women often publicly displayed the symbol of family wealth and affection on their bodies, we see in fiction that men tended to secret them. Though the novel is set in the late sixteenth century, St. Aubert's private cherishing of the miniature reflects a still common eighteenth-century practice. Frank identifies that "Men carried portraits of their lovers, wives, children, and friends in a coat, vest, or breeches pocket, at the waist in closed cases together with seals and watches, or as a pendant around their necks—but hidden under an intricately knotted cravat or other neckcloth."[15] Emily's description seems to be laced with jealousy as she incorrectly interprets her father's intimacy with the portrait's subject to have sexual associations. Melinda Rabb recognizes this collapse of the devotional and the erotic: "Human bodies are implicated in miniaturization because the body is the ultimate referent for determining scale; the intimacy requisite for experiencing the small contributes to a longstanding association between miniature and eroticism. The portrait miniature, for example, often functions as a stimulus and a locus for desire."[16] Joe Bray has argued that the revelation of the identity of the mysterious portrait miniature in *Udolpho* (as well as Radcliffe's *The Italian* [1797]) underscores its shifting associations and meanings. Bray writes, "Throughout Radcliffe's fiction, likeness is a complex, shifting quality, as tricky for the viewer to determine resemblance" (73). Yet the miniature, as in portraiture more broadly throughout the period's literature, ultimately stabilizes and its image affirms identity and proper feelings. After her father dies Emily is bequeathed his things, including the miniature, and learns that the portrait "he had so tenderly caressed" (660) is not of a former mistress as she previously intimated, but that of the Marchioness de Villeroi, St. Aubert's dead sister. When Emily sees the miniature she realizes how closely she resembles her aunt (a resemblance that helps to determine her inheritance).[17] With this realization, Emily is comforted by the bequeathed portrait as it signifies both her father's love and death, but also her birthright. In the end, the miniature confirms identity; its fluid features are replaced by static certainty.

Seen in *A Sicilian Romance* and *Udolpho*, numerous plots from the period are hinged on circulating miniatures.[18] Miniatures often play critical roles in orphan plots, so it is no surprise that in Charlotte Smith's *Emmeline, The Orphan of the Castle* (1788), miniatures of Emmeline's dead parents circulate in the novel. Easily portable, they are used to stir memories in "witnesses" who can verify Emmeline's parentage and legitimate birth. Emmeline receives a small casket from a relative that contains the "miniatures of her father and her mother, which had been drawn in Paris before her birth."[19] Having never known her mother, who died young, Emmeline cherishes her portrait. Throughout the novel the miniatures are worn, moved, placed in woven caskets, and handled. Their reemergence stirs mournful emotions, as in this occurrence when she shows her uncle his brother's miniature: "She took out the cloaths and linen, and then the two embroidered caskets, which she put on the table before her, and gazed at with melancholy pleasure, as silent memorials of her parents. . . . The sight of Mr. Mowbray's picture, which she had taken out, created in the breast of his Lordship a momentary tenderness for his niece. She had since always worn that picture about her" (341–42). Emmeline's uncle, Lord Montreville, also carries a miniature of his brother, Emmeline's father, painted when he was twenty. The narrator describes it in detail: "A likeness so striking, which he had not seen for many years, had an immediate effect upon him. His brother seemed to look at him mournfully. A melancholy cast about the eye-brows diminished the vivacity of the countenance, and the faded colour (for the picture had been painted seven and twenty years) gave it a look of languor and ill health; such perhaps as the original wore before his death, when a ruined constitution threatened him for some months, tho' his life terminated by a malignant fever in a few hours. The poor distrest Emmeline was the only memorial left of him" (149). Here the miniature takes on physical, lively characteristics—it has preserved Mr. Mowbray in a particular moment of time (of melancholy and illness). It is so accurate that when the portrait case is opened, Mr. Mowbray comes to life: he looks at his brother, casts a melancholic countenance, and embodies illness. (A portrait taken from his corpse would not have this same response.)[20] Vital is how the portrait is also linked with Emmeline's body. As living daughter, she is transformed into a material relic: "The poor distrest Emmeline was the only memorial left of him." Another miniature, this of her close friend Adelina (who is very much alive), is created by Emmeline as a gift to Godolphin, and is uniquely described as if it were a mourning portrait: "'Tis charmingly coloured; and might represent a very lovely and penitent Magdalen. The black veil, and tearful eye, are beautifully touched. But why did you indulge her in this melancholy taste?" asks Lady Clancarryl (317). Yet it is Emmeline's craftwork that vividly resembles this intersection of life and death with the relic. When Delamere reverses the miniature, he describes Emmeline's hairwork: "He looked at the back of the picture, which was of gold; and in the centre a small oval chrystal contained the words *Em. Mowbray*, in hair, and under it the name of *Adelina Trelawny*. It was indeed a memorial of Emmeline's affection to her friend; and the name was in her own hair" (317). This memorial of friendship is entwined with the physical—Emmeline's hair—which symbolically preserves her. And as a medium for inscription, her hair is also given semiotic power: transformed into writing, the hair identifies both the creator and recipient. In his discussion of relics, Laqueur notes that "The corporeal remains of the body and the other objects that the departed has left behind continue to signify" (54). Emmeline's miniature points to this ability; the dead matter speaks.

According to Marcia Pointon, portraits and landscape scenes in brooches and lockets with locks of hair mounted on the reverse are so common in Europe during the period, "whether as mourning jewelry or tokens of affection, that it is easy to overlook how invested they were in actual social relations and in representation" (61). For Lutz, Pointon, and others, it is the inclusion of hair that links the miniature to the reliquary. Pointon argues that in Protestant cultures, the "secular reliquary" that commonly combined hair and portrait "functioned in devotional ways that are self-consciously analogous to religious reliquaries" (61). Also important is the fact that bone—a common ancient relic—is a material property not only of mourning rings but of miniatures: many are composed on ivory, as we saw with the *Duchess of Gloucester* (see Figure 5.1). Kate A. Beats mentions that "Ivory was commonly used as material to paint portraits upon in England and it was used when producing funerary art," such as urns.[21] It is a permanent, "immortal" material that, like hair, represents both the immortal soul and the mortal body. Representative of imperial expansion in the East, the so-called white gold of West Africa was used in European monasteries and royal courts, often for liturgical purposes. Ivory carvings appeared on church altars and in book insets, reliquary caskets, and icons. A valuable and rare commodity, ivory would also be easily accessible to the monied classes for secular, consumer purposes in the eighteenth and nineteenth centuries. The bill-of-trade card held in the British Museum (c. 1810) of Thomas and Robert Fauntleroy, "Hard-Wood and Ivory Dealers," lists a purchase of "black ebony" and features illustrations of an elephant and a tortoise, "exotic" sources from the empire. A trade card (c. 1830) for London ivory dealer William Graham, "Ivory Bone Hardwood Turner & Worker," lists the variety of ivory items he sells, such as boxes, chess pieces, and bagatelle balls (fig. 5.2).

And in the seventeenth- and eighteenth-centuries, elephant or walrus ivory was used for fashionable portraits. Small, tangible, and skeletal, ivory simulates a talisman and is effectively transformed into another eighteenth-century form of secular relic. Similar to hair, it is an ossified element that withstands wear, immortalizes the dead, and stands in for the corporeal, for human flesh. Katherine

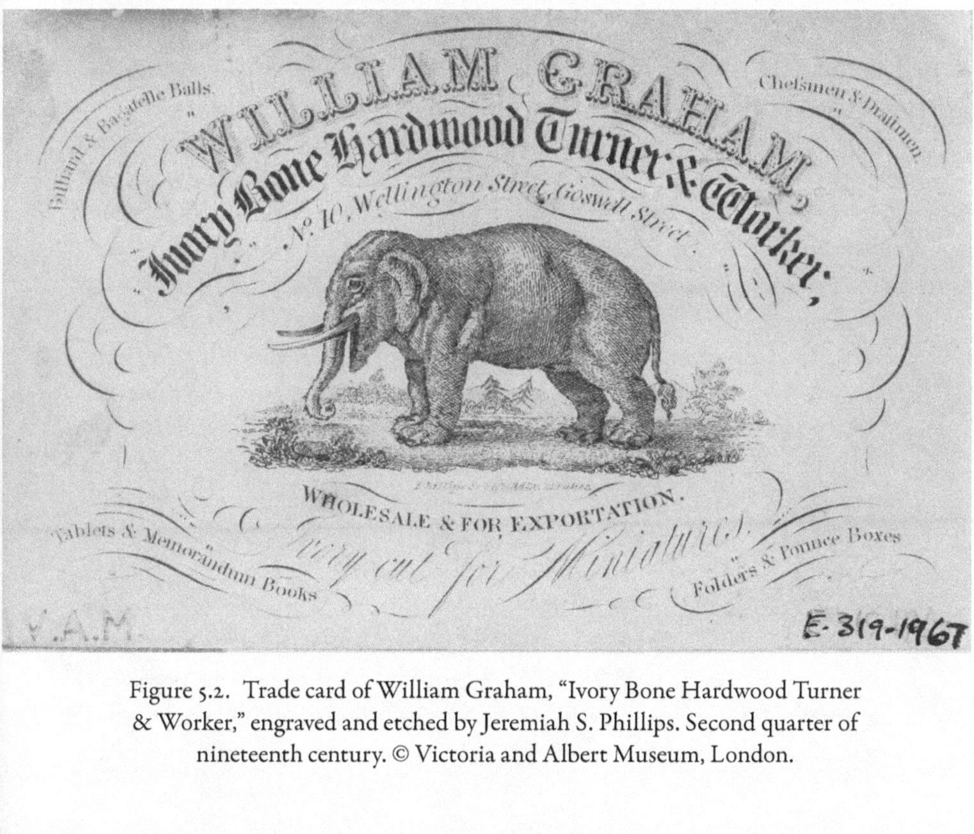

Figure 5.2. Trade card of William Graham, "Ivory Bone Hardwood Turner & Worker," engraved and etched by Jeremiah S. Phillips. Second quarter of nineteenth century. © Victoria and Albert Museum, London.

Coombs has traced the rise of the miniature on ivory to the period immediately following the exhibition of the Society for the Encouragement of Arts, Manufactures and Commerce in 1760. Here it was shown how various technical changes made it easier to paint on ivory (which, as Bray notes, had replaced vellum earlier in the century as the primary medium for the miniature [Bray 48]).[22] In his history of "white gold" and the decline of the elephant, John Frederick Walker describes this new medium for portraiture: "A thin ivory plate, no thicker than a couple of playing cards, was an ideal 'canvas' for watercolor or gouache" (92).[23] Ivory's translucence, softness, and color were considered ideal for eighteenth-century portraits, and artists like Richard Cosway often let the natural color represent the hue of the skin instead of paint (Walker 93).

While mourning practices became more aesthetic and commodified in the period, ivory increased in mourning value. Beats recognizes that ivory is a valuable material for mourning miniatures and bas-reliefs.[24] And it is an important element for a variety of other memorial gifts. While snuffboxes and toothpick boxes may appear unsentimental, they were popular love tokens in the period, and often contained portraits, precious stones, and enamel. According to a writer in *The Spectator* in 1711, boxes with miniatures were popular among military men: "I am acquainted with many a brave Fellow, who carries his Mistress in the Lid of his Snuff-box, and by that Expedient has supported himself under the Absence of a whole Campaign."[25] While consumerism may have made expensive goods easier to obtain, emotional motivations help to explain the proliferation of love tokens in the period. Rings, miniatures, hairwork, lockets, pendants, and even snuffboxes and toothpick boxes served to create a physical link between lovers, family members, and friends. (The well-known scene in Sterne's *A Sentimental Journey* [1768] where Mr. Yorick exchanges his expensive tortoiseshell snuffbox for the Franciscan monk's humble horn container reflects this sensitive, sentimental trend.) Likewise, mourning examples of these boxes soared in popularity, being used not only to commemorate the dead, but also to represent the wealth of the family in mourning. Gifted by both men and women alike, these eburnean boxes were treasured, like hairwork, for their imperishable qualities. In Chapter 3, we saw the wax bas-relief memorial of Princess Charlotte, one of numerous mourning productions commemorating her death. A Princess Charlotte enameled snuffbox with braided hair insert and miniature on ivory is a testament to the popularity of bone as a mourning material (fig. 5.3). A Georgian ivory toothpick case (c.1819) given as a mourning token reveals the portrait of Lady Cecilia Olivia Geraldine FitzGerald in its inside cover (fig. 5.4). Clearly, bone represents the renegotiation of saintly relics in the culture. While anato-

Figure 5.3. Princess Charlotte of Wales tortoiseshell gold-mounted snuffbox with braided hair insert and miniature on ivory by Richard Cosway, c. 1799. Royal Collection Trust / © Her Majesty Queen Elizabeth II 2021.

Figure 5.4. Georgian ivory toothpick case with mourning portrait inside cover, honoring the life of Lady Cecilia Olivia Geraldine FitzGerald, c. 1819. Private collection.

mists were bone collectors, and articulated and displayed morbid skeletal examples, mourners too found an appropriate form of reliquary material to handle, touch, view, carry, and exchange.

Although Pointon acknowledges that in Catholic ideology, "portrait images as such have little or no place in reliquaries," which "commonly encase—sometimes offering to view behind crystal or glass—a body part or material fragment that stands as surrogate for a presence at once mystical and phenomenological" (*Brilliant Effects* 296), she does suggest that an analogy can be made in terms of the reliquary practice, arguing that in Protestant cultures, the "secular reliquary" of portraits and hair "functioned in devotional ways that are self-consciously analogous to religious reliquaries" (296). Similar to the treatment of portraits in fictional examples, this slippage between the pious and the secular, relic and fashion, melancholy and pleasure is a common effect of devotional miniatures.

This unique response is perhaps due to the fact that portraits defy death by projecting live images of the dead. As Kamilla Elliott puts it, "[miniatures] serve as perceptual, tangible, collective memories retaining the dead in the social realism; they are material compensations for the loss of bodies."[26] A periodical writer attests in 1795, "The originals, alas! like autumnal leaves quickly perish. A portrait is the best means devised by the ingenuity of art to substantiate the fleeting form—to perpetuate the momentary existence. It is thine, O Painting! to preserve the forms which lies moldering in the tomb. . . . As the absence of the sun is supplied by artificial lights, so well-finished portraits compensate the loss sustained by the removal of the excellent originals."[27] Portraits of the dead thereby represent their subjects as eternally lost, yet their embodiment of the dead is perpetually contained and preserved. It has been well-documented that memorial miniatures such as Smith describes in *Emmeline* were in fact common in the period. They were worn as necklaces, rings, and brooches. Frank explains, "Although many mourning miniatures were copied from portraits made while the subject was alive . . . others were painted directly from the body."[28] Artists would be summoned to take a likeness of the dead, with the head and shoulders appearing on a black-bordered miniature, often with a lock of hair braided and displayed on the reverse of the portrait. Frank notes that hair was frequently "chopped up or dissolved to paint mourning miniatures . . . or simply displayed on the reverse of a portrait" (10). Unlike flesh, hair survives time and decay. "This imperviousness explains its significance in tokens of affection," Frank recognizes, "meant to outlast death, and in commemoratives" (10). Consequently, the miniature especially serves as a substitute for the body of the dead. In fictional examples, we witnessed characters transferring yearning and desire for the dead to their small portraits, carrying, speaking to, and caressing them. Their cultural purchase is marked in an article published in 1791: "The doctrine of Talismans . . . has stood its ground. . . . Mourning rings, miniature pictures, lockets, devices, armorial bearings are all on this principal."[29] An important visual example from the period attests to the emotional value of miniatures. In the full-size portrait *Margaret Cocks* (c. 1787) by Richard Cosway, arguably the most important miniature artist of the period, Margaret is depicted in a melancholic pose, eyes raised to heaven, with pastoral imagery behind her (fig. 5.5). Longing for her sister Mrs. Mary Russell, who died when Margaret was thirteen, Margaret cherishes her miniature.[30] Margaret's wearing of her sister's miniature on a chain around her waist invokes a form of secular memorial. Though uncertain, we can speculate that the letter she is reading was written by Mary, and Cosway emphasizes the significance of the memorial relic to Margaret's melancholic contemplation.

A similar sentimental "double portrait" is described by Mary Shelley in *Frankenstein* (1818), which is embedded with numerous mourning artifacts (and of course the creature's body itself is the most powerful artifact comprising human remains and decomposing flesh). The valuable miniature of the deceased Caroline Beaufort is carried by her son William as a mourning portrait. Symbolizing maternal death and loss, the miniature's depiction of a beautiful woman launches the creature's murderous rampage when he sees it worn by the child: "As I fixed my eyes on the child, I saw something glittering on his breast. I took it; it was a portrait of a most lovely woman. In spite of my malignity, it softened and attracted me. For a few moments I gazed with delight on her dark eyes, fringed by deep lashes, and her lovely lips; but presently my rage returned; I remembered that I was forever deprived of the delights that such beautiful creatures could bestow and that she whose resemblance I contemplated would, in regarding me, have changed that air of divine benignity to one expressive of disgust and affright."[31] Hauntingly descriptive and detailed (amazingly, he can see her "deep lashes"), Caroline's miniature, as with many I have discussed, animates when viewed. Her image calms the creature, and draws him into her beautiful depiction, launching his fantasy of the "delights" she could bestow, then just as quickly her air of "divine benignity" in his imagination switches to "disgust and affright" as she reverses the power of the gaze and regards the viewer. Most strikingly sentimental (and in certain ways disturbing to a modern reader) is the full-size portrait of Caroline mourning her father, commissioned by her husband (Victor's father, close friend of the departed). Following his mother's death and his brother William's murder, Victor turns to the painting: "I gazed on the picture of my mother, which stood over the mantelpiece. It was an historical subject, painted

Figure 5.5. Richard Cosway, *Margaret Cocks, later Margaret Smith*, c. 1787. Oil on canvas. 36 × 27¾ in. Courtesy of the Huntington Library, Art Museum, and Botanical Gardens. Gift of Mr. and Mrs. Clifford E. Clinton (60.8).

at my father's desire and represented Caroline Beaufort in an agony of despair, kneeling by the coffin of her dead father. Her garb was rustic, and her cheek pale; but there was an air of dignity and beauty, that hardly permitted the sentiment of pity. Below this picture was a miniature of William; and my tears flowed when I looked upon it."[32] Here it is not his mother's mourning portrait that inspires Victor's tears, though, but the more intimate miniature of William. The scene demonstrates the growing sentimentality toward death and the power of the miniature to immortalize the missing body. As we have seen, portraits have a spectral power to conjure the dead and deep emotion. Like so many of the novel's unresolved mysteries and loose ends, Caroline's and William's miniatures are eternally suspended by Shelley in *Frankenstein*—they are never bequeathed or claimed.

While hair is the most popular human remain with which to compose artifacts, we saw in Chapter 2 that heart and viscera burial continued to be a part of mourning sentiment in the period. Heart caskets and organ/viscera urns, though not common, were commissioned well into the nineteenth century. Neoclassical urns were necessary funerary accessories for the distinguished and wealthy, whether housing human remains (lungs, heart, viscera, ashes) or symbolically invoking them. The mourning woman with the urn is a logical expression of grief across history. While Beats and others see the shift from graphic illustrations of skeletons and the decaying body as representing "an emphasis on beauty" that "did not focus upon the dead body,"[33] many of these images offered a melancholic reassurance that turned attention to not only the preserved dead body, but to the experience of mourning that involved direct contact with the dead. This is depicted in several portraits in the period, including mourning miniatures. Earlier in this chapter. I discussed Cosway's full-size mourning portrait of Margaret Cocks holding her sister's miniature (fig. 5.5). Cosway's miniature *Portrait of Margaret Cocks, mourning her sister's remains* (c. 1787) is noticeably different for it incorporates the neoclassical iconography of an urn symbolizing her sister Mrs. Mary Russell's remains (the urn is inscribed "M.R./1786") (fig. 5.6). George Lumley also executes a sentimental double portrait in *The Lady Mary Fenwick* (c. 1737), held at the British Museum. The mezzotint etching of the widow ("relict") depicts her in mourning, looking in the distance, holding the miniature of her husband, Sir John Fenwick, in her right hand. This portrait-within-a-portrait is a relatively rare example of the sitter holding the "body" of the deceased.[34] (In Chapter 3, we saw a similar example in Zoffany's portrait of Lady Salusbury holding the will of her husband with black mourning seal.) Behind Lady Fenwick is an uninscribed urn, symbolizing Sir John's remains. Like the depiction of the urn in the portrait of Lady Fenwick, perhaps for the eighteenth-century viewer the urn in Cosway's miniature invokes viscera preservation. Cosway's neoclassical reference, though, most likely symbolizes a cinerary urn and the ancient practice of cremation (cremation was not allowed by the Church of England during the period). Margaret's clutching and embracing of the urn ("mourning her sister's remains") powerfully signals the sentimental desire for physical contact with the dead.

Figure 5.6. Richard Cosway, *Margaret Cocks, mourning her sister's remains* (formerly called Mary Russell mourning her mother's remains), c. 1787. Portrait miniature. Watercolor on ivory. 72 × 57 mm. © Victoria and Albert Museum, London.

Skeletal Fictions

Previous chapters have collated various actual and fictional examples of the circulating dead and dead-like body: dissected, embalmed, displayed, exhumed, preserved in spirits, molded from wax, composed of hair. I wish to close this chapter with examples of another form of circulating relic: skeletal remains.

Clara Reeve's *The Old English Baron* (1778) is set in medieval England, its plot centered on the true parentage of Edmund Twyford, the supposed son of a peasant. Having been brought to live with the Baron Fitz-Owen, Edmund is befriended by Sir Philip Harclay, who is struck by his resemblance to his lost friend Lord Arthur Lovel. While staying in the haunted wing of Lovel Castle (the seat of Baron Fitz-Owen), Harclay and Edmund discover a bloodstained suit of armor and the portraits of Lord and Lady Lovel, evidence that helps them start to piece together Edmund's true origins and the fate of his parents. Eventually, Edmund discovers he is indeed the son of Lord Lovel, who was murdered by Sir Walter Lovel (as he would gain the estate). Yet final proof of his parents' remains is needed to affirm Edmund's birthright and entitlement. The narrator explains: "Here is a young man, supposed the son of a peasant, who, by a train of circumstances that could not have happened by human contrivance, discovers not only who were his real parents, but that they came to untimely deaths. He even discovers the different places where their bones are buried, both out of consecrated ground, and appeals to their ashes for the truth of his pretensions. He has also living proofs to offer, that will convince the most incredulous."[35] Reeve then describes what will be a foundational scene for the early Gothic novel:

> Edmund bade them bring shovels and remove the earth. While they were gone, he desired Oswald to repeat all that passed the night they sat up together in that apartment, which he did till the servants returned. They threw out the earth, while the by-standers in solemn silence waited the event. After some time and labour they struck against something. They proceeded till they discovered a large trunk, which with some difficulty they drew out. It had been corded round, but the cords were rotted to dust. They opened it, and found a skeleton which appeared to have been tied neck and heels together, and forced into the trunk. "Behold," said Edmund, "the bones of him to whom I owe my birth!" The priest from Lord Graham's advanced. "This is undoubtedly the body of the Lord Lovel; I heard his kinsman confess the manner in which he was interred. Let this awful spectacle be a lesson to all present, that though wickedness may triumph for a season, a day of retribution will come!" (116)

Later, Walter testifies that out of jealousy he had Arthur murdered and his corpse brought back to the castle, tied "neck and heels," and placed in a trunk under the floorboards. This unearthing of the trunk and discovery of his father's skeleton is both an "awful spectacle" and a moment of sentiment: "The bones of him to whom I owe my birth!" The unearthed bones are also the physical evidence of both horrific murder and identity and birthright: Edmund is indeed the noble peasant who discovers his aristocratic origins and regains his estate.

Longing to have his parents' remains rejoined and properly buried, Edmund next plans to locate his mother's bones:

> "The first thing I propose to do," said he, "is to have a coffin made for these honoured remains. I trust to find the bones of my other parent, and to inter them all together in consecrated ground. Unfortunate pair! you shall at last rest together! your son shall pay the last duties to your ashes!"
>
> He stopped to shed tears, and none present but paid this tribute to their misfortunes. Edmund recovered his voice and proceeded. "My next request is, that Father Oswald and this reverend father, with whoever else the gentlemen shall appoint, will send for Andrew and Margery Twyford, and examine them concerning the circumstances of my birth, and the death and burial of my unfortunate mother." (116)

Edmund's initial response to human remains is not one of Gothic horror (such as Radcliffe depicts in *Udolpho* and *Romance of the Forest* [1791]). The skeletons of his parents are thought of in sentimental terms and with emotional gestures. In fact, the Gothic enjoys this slippage between the horror of decaying remains of the anonymous and the revered relics of loved ones.[36]

Next, Edmund upholds the honor and dignity of his parents' bones, ensuring they are housed together in a "stately coffin": "The next day Andrew directed them to the place where the Lady Lovel was buried, between two trees which he had marked for a memorial. They collected the bones and carried them to the Castle, where Edmund caused a stately coffin to be made for the remains of the unfortunate pair. The two priests obtained leave to look in the coffin buried in the church, and found nothing but stones and earth in it. The commissioners then declared they were fully satisfied of the reality of Edmund's pretensions" (117). Knowing the townspeople are concerned that the "poor Lady had not a Christian burial," "Edmund then desired father Oswald to give notice to the friars of the monastery of St. Austin, that with their permission the funeral should be solemnized there, and the bones interred in the church" (128): "Preparations being made for the funeral, it was performed a few days after. Edmund attended in person as chief mourner, Sir Philip Harclay as the second; Joseph desired he might assist as servant to the deceased. They were followed by most people of the village. The story was now become public, and every one blessed Edmund for the piety and devotion with which he performed the last duties to his parents.— Edmund appeared in deep mourning; the week after, he assisted at a mass for the repose of the deceased." Sir Philip Harclay then orders a monument to be erected to the memory of his friends, with the following inscription:

> "Praye for the soules of Arthur Lord Lovele and Marie his wife, who were cut off in the flowere of theire youthe, by the trecherye and crueltie of theire neare kinnesmanne. Edmunde theire onlie sonne, one and twentie yeares after theire deathe, by the direction of heavene, made the discoverye of the mannere of theire deathe, and at the same time proved his owne birthe. He collected theire bones together, and interred them in this place: A warning and proofe to late posteritie, of the justice of Providence, and the certaintie of Retribution." (128)

In the end, Edmund's identity and birthright are confirmed, in large part, by skeletal evidence. Locating his parents' remains and having them buried together signifies Edmund's parentage and legacy. The tombstone's inscription "collected theire bones together" points to the endurance of the bones which are both "a warning and proofe" of justice and providence. But as entwined relics, the bones are also given aesthetic and emotional purpose and symbology. Outside of a commodified or anatomical setting, bones are sought after, cherished, and lovingly intermingled.

This chapter's last fictional example also involves circulating relics of the past: a dusty manuscript, a rusty dagger, a signet seal, and as in Reeve's novel, skeletal remains hidden in a trunk. Set in Roman Catholic Europe, Radcliffe's *The Romance of the Forest* is centered on the heroine Adeline, who has been abandoned by her father and placed under the protection of the La Motte family, who are fleeing Paris. They take refuge in a ruined abbey in southern France, where the relics are discovered by La Motte in a secret passage "leading to the subterranean cells" that contains a hidden chest:

> Lifting the lid, he saw the remains of a human skeleton. Horror struck upon his heart, and he involuntarily stepped back. During a pause of some moments, his first emotions subsided. That thrilling curiosity, which objects of terror often excite in the human mind, impelled him to take a second view of this dismal spectacle.
>
> La Motte stood motionless as he gazed; the object before him seemed to confirm the report that some person had formerly been murdered in the abbey.[37]

Afraid of the "dismal spectacle," La Motte avoids the secret chamber: "He was anxious that his family should not perceive the skeleton; an object, which would, probably, excite a degree of horror not to be overcome during their stay" (54). Similar to the mystery behind the black veil in *Udolpho*, Radcliffe delays the revelation of the skeleton's history for hundreds of pages; the horrific bones are kept hidden. In a different chamber in the abbey, Adeline discovers a hidden chest that holds a rusty dagger and a manuscript: "It was a small roll of paper, tied with a string, and covered with dust" (116). Secreting it to her own chamber, she reads the narrative, which details a man's imprisonment and torture in the abbey, then returns it to the chest. Similar to the treatment of the skeleton, the contents of the chest remain secret and buried in the depths of the ancient abbey.

By the novel's close they are unearthed, and we learn that the remains are of Adeline's father, the marquis, who had been imprisoned in the abbey and murdered years ago. Two hundred and fifty pages later Adeline recalls her discovery of the manuscript: "When she remembered the manuscript so singularly found, and considered that when she wept to the sufferings it described, her tears had flowed for those of her father, her emotion cannot be easily imagined. The circumstances attending the discovery of these papers no longer appeared to be a work of chance, but of a Power whose designs are great and just, 'O my father!', she would exclaim, 'your last wish is fulfilled—the pitying heart you wished might trace your sufferings shall avenge them'" (346). The dusty manuscript resurrects Adeline's father; she weeps for and speaks to him. Following the execution of d'Aunoy, her father's murderer, Adeline has her father's skeletal remains (no longer a source of Gothic horror) properly buried: "They were deposited with the solemn funeral pomp becoming his rank: Adeline attended as chief mourner" (355). Initially a source of Gothic horror, her father's remains are reconfigured as sentimental relic.[38]

Instead of preserving a relic of ash and bone, Adeline treats his manuscript ("covered with dust") as a representation of the decayed body: "The sufferings of her dead father, such as she had read them recorded by his *own hand*, pressed most forcibly to her thoughts. The narrative had so much affected her heart, and interested her imagination, that her memory now faithfully reflected each particular circumstance there disclosed. But when she considered that she had been in the very chamber where her parent had suffered, where even his life had been sacrificed, and that she had probably seen the very dagger, seen it stained with rust, the rust of blood! by which he had fallen, the anguish and the horror of her mind defied all control" (346–47). Again, the manuscript seems to have the power to resurrect her father: his dead hand "pressed most forcibly to her thoughts," and she imagines his physical torture, blood, and murder. Adeline's realization that she was present in his "tomb" and had seen his blood reinforces these corporeal references, inspiring her to preserve and cherish the physical representation of his body: "The MS. that recorded his sufferings had been found at the Abbey, and delivered to her by M. V—, and she preserved it with the pious enthusiasm so sacred a relique deserved" (355). With this, Radcliffe in fact defines a form of secular relic: the manuscript having associations with Catholic piety ("pious enthusiasm"), yet preserved (and therefore readily accessible) for the pleasure of the individual mourner. This morbid preservation of the written word is apt for a novel that mentions "mourning" and "mournful" twenty-two times and that quotes graveyard poetry throughout. Significant is that Radcliffe directly aligns the dusty manuscript and "relique" with fiction making: the dead hand of narrative is preserved for posterity.

In addition to her father's testament of torture, the novel's other mysteries are unveiled with the story told by one last object. Throughout the novel the orphaned Adeline covets a family seal as a type of strange secular relic. In fact, it acts as a form of DNA, verifying her true identity. Bearing the arms of her mother's family, a note written by Adeline is purloined by the conniving marquis, who recognizes the seal and its evidentiary role in his ultimate fall: "As the history of the seal which revealed the birth of Adeline is rather remarkable, it may not be amiss to mention, that it was stolen from the Marquis, together with a gold watch by Jean d'Aunoy: the watch was disposed of, but the seal had been kept as a pretty trinket by his wife, and at her death went with Adeline among her clothes to the convent. Adeline had carefully preserved it, because it had belonged to the woman whom she believed to have been her mother" (345). With this "bodily" evidence, the Marquis de Montalt's fate is "sealed"; on the revelation that he arranged the murder of his half-brother, Adeline's father, he is imprisoned and commits suicide by poison. Noted in Chapter 3, wax mourning seals were material representations of death and mourning; the seal symbolizes familial status but also the body and corpse (fig. 5.7). Adeline covets and "preserve[s]" the maternal signet that creates the seal. With this, Radcliffe centers the key to Adeline's birthright on a diminutive yet symbolically powerful memorial artifact.

Articulated Bones

The coexistence of bone as Gothic horror, material for mourning objects, valuable commodity, and anatomical display speaks to its fungible purpose in fiction. And as we have seen, bone can easily transform into secular, sentimentalized relic, a more acceptable function from its previous purpose as Catholic reliquary. This

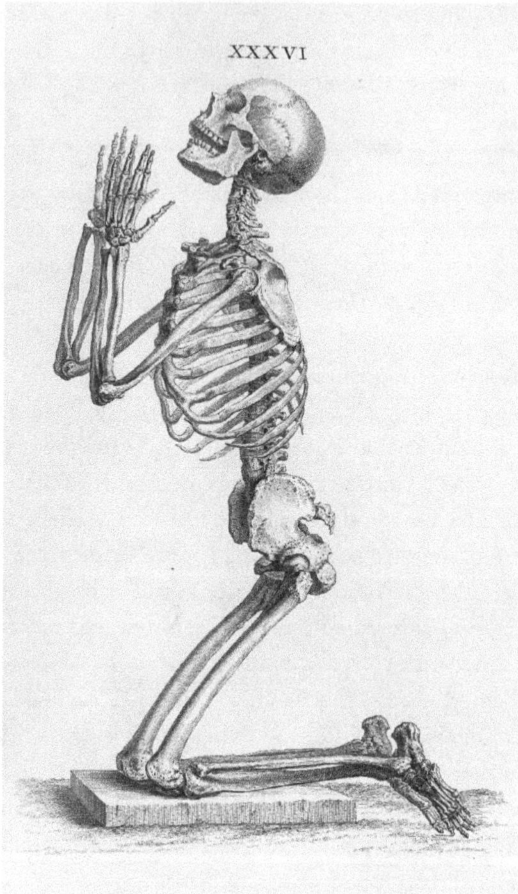

Figure 5.7. William Cheselden's *Osteographia, or the Anatomy of the Bones*, Plate 36, 1733. Courtesy of the U.S. National Library of Medicine.

sentimentality even extended to adored pets in the late eighteenth century (a return to ancient burial practices, reflecting the Egyptians). Sculptor Anne Damer (d. 1828), famous for her sculptures of pets, especially dogs, directed in her will to be buried not only with her sculpting tools and apron, but with the bones of her favorite dog, Fidèle, who had died many years before in Portugal.[39]

Human anatomy was a necessary part of public and private medical school education in the eighteenth century, and the demand for anatomical specimens was keen. Anatomy lectures began with a skeletal presentation of the body, and treatises on osteology instructed readers on the proper construction of skeletons.

Guerrini remarks, "A good human skeleton was a valuable commodity."[40] Subsequently, the period saw several publications depicting articulated bones. Surgeon William Cheselden's *Osteographia, or the Anatomy of the Bones* (1733) is a remarkable example of the articulators' art: skeletal animals are posed as if active and in motion, as are human skeletons (including that of women and children).[41] The use of the camera obscura, copperplate engravings, and life-like vignettes designate the book's importance in the history of anatomical illustration. For example, Plate 36 (fig. 5.8) depicts a side view of a male skeleton praying. With the rising interest in anatomy, sellers of human bones and fully articulated human and animal skeletons (typically for veterinary college purposes) therefore found their wares in high demand. The trade card (c. 1791) for one such London skeleton dealer, Nath Longbottom (fig. 5.9), advertises that he

> Sells Skeletons of different sizes & both sexes, of good color & accurately articulated, & packs them safe either for Sea or Land carriage. He also mounts for such gentle-men as have loose sets of bones.

Simon Chaplin cites an eighteenth-century midwifery lecturer who promises his students "a great variety of Preparations, such as skeletons of different makes, Pelvis's

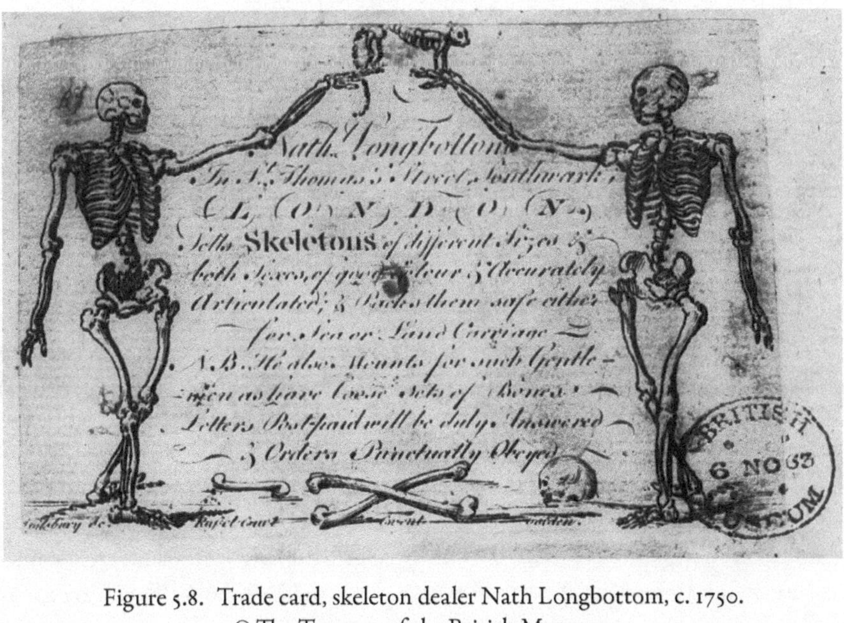

Figure 5.8. Trade card, skeleton dealer Nath Longbottom, c. 1750.

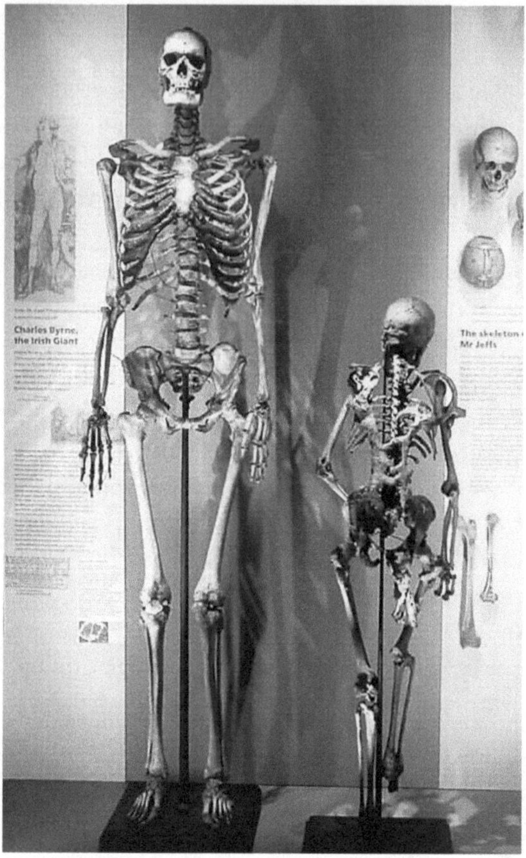

Figure 5.9. Charles Byrne's skeleton (d. 1783; *left*), c. 2017. Hunterian Museum, Royal College of Surgeons of London. Photograph reproduced with the permission of Emőke Dénes. Creative Commons.

of diff't diameters & depths . . . Foetal skulls and skeletons with remarks on them & the structure of the pelvis."[42] In one account of the public dissection of a murderer's corpse at Surgeons' Hall (which women could indeed attend), a newspaper report describes that when a skeleton displayed in the hall fell to the ground, "the women fainted, and the men were frightened."[43] David Jones has shown that popular phantasmagoric shows from the period included the transformation of a body into a skeleton, achieved through a doubling of glass slides. Audiences in France were treated to the sight of "The Three Graces, turning into skeletons," and a London show in 1807 depicted Louis XVI and Benjamin Franklin undergoing a similar transformation.[44] The skeleton was thereby an element and source of spectacle making. The articulated skeleton of notorious "Lost Property" dealer, "Thief-Taker General," and criminal gang leader Jonathan Wild (currently displayed in the Hunterian Museum at the Royal College of Surgeons, England) stands as an example of the intersection of science and performance. Following Wild's well-publicized and attended execution in 1725, his corpse was exhumed by grave robbers and sold to the Royal College of Surgeons for dissection, the anatomists later wiring his skeleton for permanent display.[45] These various examples attest to this simultaneous register of morbid sentimentalizing, anatomical display and inquiry, and Gothic, pleasurable horror, signaling the eighteenth-century transformation of the skeleton's former role as religious *memento mori*.

The anatomists' interests in the human skeleton may have been seen as purely scientific, but as Guerrini reminds us, "they were highly aware of its religious and symbolic meanings."[46] Noted earlier, skeletal iconography was frequently used as *memento mori* in the previous century, and skeletons still remained familiar icons in Catholic countries. Examined throughout this project, though, the human skeleton decreased in popularity as a symbol of mourning in the period. As morbid icons of *danse macabre* and *vanitas* art were no longer popular in the culture, human skeletons were symbolically refashioned into ivory secular memorials, materials for miniatures, snuffboxes, and other mourning gifts that instead depicted willows, urns, and weeping mourners.

While skeletons were collected for articulation and the interests of anatomical science, corpse medicine became an element of eighteenth-century medical practice. Perhaps shocking to modern understanding, ground human skull was a component of apothecary shops, consistently prescribed for epilepsy and other ailments in the eighteenth century. Ancient mummy was also ingested for medicinal purposes (including its supposed aphrodisiacal powers). This is evidenced in numerous publications, including Samuel Johnson's *Dictionary*, which lists "mummy" as a medicine.[47] And while texts, illustrations, public exhibitions, anatomy schools, and private collections displayed the human skeleton, there were also fascinating cases of "celebrity" skeletons. For example, in "'Alas Poor Yorick!': Jonathan Swift, Madness and Fashionable Science," Helen Deutsch recounts the early twentieth-century phrenological study of writer Jonathan Swift's skull, and the medical inferences about his disease and death.[48] In *The Corpse: A History*, Christine Quigley observes that exhumations of the famous were often committed to determine if a person existed at all, as in the case of famed Robin Hood companion Little John. In the eighteenth century (about 1785) an ancient grave in Hathersage, Derbyshire, belonging to a John Little was opened: "A thigh bone contained in the grave indicated the occupant's height to be seven feet tall, the conclusion being that the remains were those of 'Little John'" (140).[49] Noted in the previous chapter, the late eighteenth-century exhibition and dissemination of Milton's corpse and relics (including hair and bone) are sensationalized in Philip Neve's *Narrative of the Disinterment of Milton's Coffin* (1790). The Carl H. Pforzheimer Collection of Shelley and His Circle in the New York Public Library houses what are purportedly remnants of Percy Shelley's well-traveled skull.[50] Seemingly, the late eighteenth century experienced numerous accounts of revered literary bones (and even kidney and bladder stones),[51] apt for a culture seeking new ways to memorialize.

Chapter 2 of this book concludes with the eighteenth-century circulation of Oliver Cromwell's embalmed head, and I think it apt to close this chapter with two similarly unique, ossuary, posthumous narratives: those of Charles Byrne and Laurence Sterne. Both of their corpses were victims of anatomists and are shrouded in mystery. And similar to the skeletal remains in Reeve's and Radcliffe's works, theirs would not be properly buried for many years (if at all in Byrne's case).

Joshua Reynolds's famous portrait of the anatomist and morbid collector John Hunter contains the skeletal feet of Patrick O'Brien (or Charles Byrne) (see Figure 2.14), the celebrated "Irish Giant" who stood at around seven feet, seven inches. Though he promoted himself as a "freak" and charged audiences to see him, Byrne was terrified of being dissected, articulated, and displayed, and it was rumored he had a will that stated his body would not fall into the hands of an anatomist. Despite making arrangements with friends to bury him at sea, on Byrne's death in 1783 an undertaker was bribed and his body was removed from the coffin and secretly sold to the highest bidder, John Hunter, who had made it no secret that he would covet the articulated Byrne. For the next several years, speculation circulated in the press about what had supposedly been a failed attempt to bury Byrne at sea, and the great value his skeleton would have for a collector. Ruth Richardson remarks that now Charles Byrne's "skeleton greets one on entering the Hunterian Museum, Royal College of Surgeons of London, and for whose corpse the anatomist John Hunter paid over £500."[52] To avoid detection, Hunter had chopped the body into pieces and boiled it down to the bones, waiting four years to admit owning the skeleton before displaying it at the Hunterian, where it has indeed been displayed for over two hundred years despite Byrne's fears and wishes. Incidentally, there is a current campaign to stop the display of and to properly bury Byrne. Perhaps after all of these years welcoming visitors to the Hunterian, Byrne's skeleton will rest in peace.

As his Yoricks defy the grave, novelist and clergyman Laurence Sterne also experienced posthumous notoriety, that is, speculation regarding the unearthing of Sterne's skeleton and the journey of Sterne's skull, a more than uncanny tribute to Sterne's (and Shakespeare's) variations of Yorick. While several scholars have traced the history of Sterne's death, burial, and disinterring, and the reinterment of his skull in 1969, I am interested in its eighteenth-century circulation and narrative.[53] For my discussion, his traveling bones signal both the rise of the anatomist and the cultural desire for memorialization and secular reliquary. Dying on March 18, 1768, Sterne would be rumored to have died in hired lodgings, robbed by his attendants "of his gold sleeve-buttons" (Ferriar, *Illustrations of Sterne* from 1798), and subsequently disinterred and "carried to Oxford as a subject for the Lectures of the Anatomical Professor in that University." In 1769 the *Public Advertiser* jests, "It has been whispered about some Time in great Confidence, that the Skeleton of the famous Yorick has been exhibited in one of our English universities, and it seems now to be put beyond all Doubt by a Gentleman's having applied in Town to search for the Body, and it could not be found. Another Gentleman is well assured of the Identity of his Skull by two or three Teeth being remarkably prominent, which were well remembered by those who knew the Deceased."[54] Buried at St. George's Hanover Square, Sterne was potentially one of many who were disinterred by grave robbers. As Kenneth Monkman, W. G. Day, and other scholars detail, publications throughout the eighteenth century speculated about the location of Sterne's body, and even questioned whether he was buried at St. George's. Some recorded that Sterne was disinterred and anatomized at Cambridge. Others noted that witnesses were "present at the dissection" and "recognized Sterne's face the moment [they] saw the body."[55] Recent biographer Ian Campbell Ross supports this theory, stating "Interred in St George's burial ground in London on March 22, he was surreptitiously resurrected, appearing—to the astonishment of one onlooker who had known him in life—on the anatomist's table at his old university at Cambridge. Hastily returned to the cemetery, Sterne's corpse lay there until the burial ground was threatened by redevelopment in 1969."[56] Since there was no documentation about Sterne's original burial, in 1769 an erroneous tombstone was erected "Near to" the place where Sterne had supposedly been buried. Indeed, over the next two hundred years, misinformation surrounding the circumstances of Sterne's death and his resting place would be disseminated. Was he indeed anatomized? If so, would he be reinterred in the same burial ground and grave? On exhumation of St. George's decrepit burial ground in 1969, Monkman notes that "Of the 11,500 skulls exhumed . . . only two or three had been sawn" (65). Experts agreed that

Sterne had indeed been a victim of the resurrectionists, thereby his skull would reveal an anatomist's craniotomy (fig. 5.10). Determining one of the sawn skulls to be Sterne's (using measurements from busts, medical examination of the skull, and features), Monkman and his fellow Shandeans reinterred the skull (along with a femur and other bone fragments) in 1969 at St. Michael's, Coxwold, near Shandy Hall. Sterne's ossuary journey had finally come to an end.

In the eighteenth century, *memento mori* rose in popularity as they transformed into miniature, aesthetic, sentimental, secular objects. Simultaneously, *memento mori* were transformed into objects of curiosity, displayed in cabinets, pickled in jars, dissected and preserved by anatomists. These last two chapters in particular have examined how sentimental mourning practices intimately materialize in fiction. As previously argued, I see funerary customs influenced by the growing circulation of human specimens in the period. And as dissection was integral to Enlightenment anatomy, individual control over the corpse was of growing concern. As a result, the corpse becomes a distinctive form of material culture, and how it is disposed and commemorated of great importance. Exposing insights that history alone has not yet provided, *Death and the Body in the Eighteenth-Century Novel* has demonstrated how practices and attitudes in fiction uniquely reflect the culture's engagement with the body and human remains.

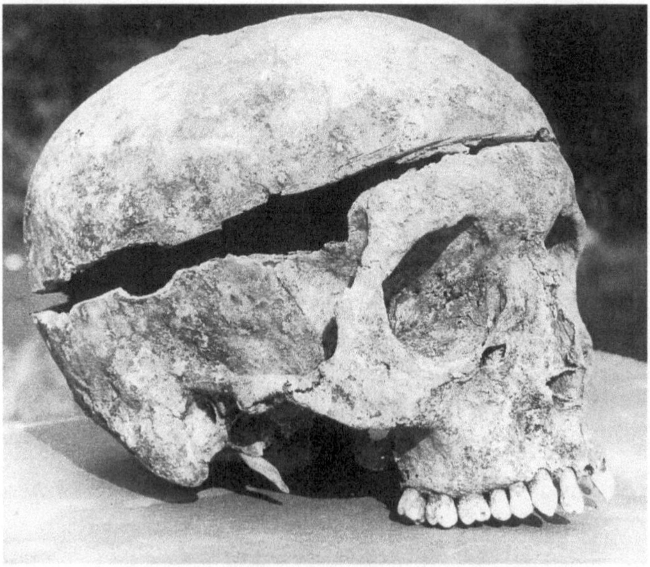

Figure 5.10. Laurence Sterne's skull (d. 1768), c. 1998. Photograph courtesy of the International Laurence Sterne Foundation.

Mourning objects prompt the understanding that fiction emerges in intermedial spaces (visual, textual, material) as well as representational. Gathering extensive archival examples, *Death and the Body* has claimed in part that a material, corporeal, sentimental, commodified death culture did not emerge in the Victorian era, and that much of this cultural evidence is embedded and archived in the British novel. And as *Clarissa* and so many other novels attest to, it claims that individuals sought ways to sentimentally preserve their bodies and posthumous state, finding comfort in memorializing elements such as hair and bone that cannot decompose. I think it is this very resistance to decomposition—the attempt to immortalize—that in fact underlies the purpose of these tokens in fiction, which at its foundation eternally seeks to compose.

AFTERWORD

Many eighteenth-century relic and mourning practices became widespread and sentimental in the Romantic period, later culminating in "the Victorian celebration of death" with Queen Victoria, "the Widow of Windsor," at the helm. While Victorians sought evidence of life after death, the cult of mourning provided the materialistic displays (and extremes) of remembrance. Elaborate mourning adornments all emerged as public signs of loss.[1] In *Death and the Future Life in Victorian Literature and Theology*, Michael Wheeler remarks that during "the Victorian Age, highly conventionalized social customs and funerary rituals eased the transition from the deathbed to the bed that is the grave."[2] And of course the body itself became the most important relic. In fact, the Victorians fetishized the corpse, creating material objects that were effigies not only *for* the dead but *of* the dead—such as mourning cards, deathbed portraits and photographs, and hair lockets and artwork. Secular rituals involved the washing, watching, waking, and viewing of the corpse. Ruth Richardson recognizes that at the time of the 1832 Anatomy Act (which allowed for the corpses of the poor to be used for dissection purposes), there was a prevailing belief in the existence of a strong tie between body and soul for a period of time after death: "The result was an uncertain balance between solicitude toward the corpse and fear of it."[3] Whether or not physical resurrection would actually take place, the protection of the identity and integrity of the corpse in the nineteenth century served as a metaphor for the possibility of providing a secure future for the soul.

Though by no means a homogeneous experience, the Victorian cult of death was an undeniable sociohistorical and psychological response of a bereaved culture.[4] A definitive reasoning most likely to never be agreed on, we can at least determine that a combination of economic, psychological, and consumer interests spurred the Victorian cult of mourning. David Cannadine notes, "It is arguable that the Victorian celebration of death was not so much a golden age of effective psychological support as a bonanza of commercial exploitation."[5] Pat Jalland similarly questions the widespread consumer culture surrounding mortality, where the existence of mourning teapots and pincushions, and spectacular, extravagant funerals, signifies the expansion and professionalization of death's commodification.[6] The value placed on sentiment and the individual witnessed in the eighteenth century had intensified the demand for commemoration. Growing secularization and an expanding middle class, "keen to display their wealth and status through conspicuous consumption, fuelled the demand for death-related goods and services," Claire Wood observes.[7] As the industry of death and mourning proliferated, the experiences of dying and mourning became public signs of material survival and moments of cultural fascination shared in various forms and levels of expense by all classes. "All this enthusiasm might be considered purely an interest in social display," John Kucich explains, "a means for the middle class to prove both wealth and gentility, were it not for the spilling over of the phenomenon into literary taste."[8] Victorian sentiment thus conceived of death as a fetishized event, one persistently dramatized throughout the century.

For these reasons, Victorian literature appears to be obsessed with and even to eroticize death, taking examples from *Clarissa* and refashioning them for a more modern, diverse culture reacting against Romantic aesthetics. As a result, Victorian novels are full of lingering illnesses and exquisite corpses: both the virtuous (such as Charles Dickens's Little Nell from *The Old Curiosity Shop*) and the sexual transgressor (such as Elizabeth Gaskell's Ruth) can beautifully endure their illnesses.

Literary depictions of death proved important to the social and cultural understanding of absence, separation, and displacement in an ever-increasing chaotic and dismembered world. As Queen Victoria retreated into perpetual mourning, finding solace in elegies such as Tennyson's *In Memoriam*, and as artists and writers creatively indulged their own mournful experiences, Victorians appropriated the previous century's sensibilities, constructing a fascinating cult of mourning. This interest would spill over into occultish practices (séances, table rapping, spirit photography, and so on). Deborah Lutz notes that as spirit photographs were produced in large numbers between 1880 and 1920, they can be seen "as another transitional-period flourishing—a final flaring up on the edge of extinction."[9] Many technologies (telegraph, phonograph, radio) that emerged in rapid succession during the late nineteenth and early twentieth centuries thereby hauntingly preserved the voice, the spirit, and the image of the body. As a result, relics were displaced and replaced, later serving as embodied evidence of eccentric (today, perhaps even seen as creepy) practices.

As the acceptance of preserved remains peaks in the period, it signals its own demise. Inevitably, technology mechanizes mourning and the body, and the pomp and circumstance of Victorian funeral pageantry implodes. In part, cherishing remains takes on a horrid, monstrous connotation in light of the mutilating effects of industry and war. By the end of the nineteenth century, we see another shift in attitudes about the corpse and death: they are something to be feared, denied, and not embraced. It is fitting that at the *fin de siècle*, horrific and Gothic scenes of death will dramatically contrast the Victorian ideal of a beautiful, Christian death. In fact, the ugly death of the vampire and the gruesome corpse of punished villains emerge as powerful literary emblems for a culture mourning the loss of certainty and tradition, coveting yet fearing modernity. Lutz reminds us that "Mobilization and ungroundedness, coupled with the secularization and medicalization of death, caused the embodied keepsakes of Victorians to be relegated to the periphery, to become the practice of eccentrics or the religious minority."[10] In this sense, we may never return to embracing, commodifying, or aesthetically preserving the dead. And most likely Neo-Victorian and Gothic subcultures may rest assured that their aesthetics will remain fashionably in the margins.

Notoriously peculiar, Victorian death rituals were often shrouded in the mystical and erotic. They offered passage from life to death, practical therapeutics, and rites for the living to ease pain and in some ways extend grief. Today we may see this necroculture as obsessive and unhealthy, too funereal and melancholic. Yet our violent shift to the cold and distracted approach to loss and mourning inevitably has its own repercussions. Without a collective observance of death in the West, and amid a current global pandemic with daily reminders of loss, perhaps we can pivot our gaze to the past in order to recognize the emotional value of sentimental memorialization. Through examining narratives that encapsulate dying, death, and mourning artifacts, we can trace the eighteenth century's impactful aesthetic legacy.

NOTES

INTRODUCTION

1. Some noteworthy studies include Paul S. Fritz's "The Undertaking Trade in England: Its Origins and Early Development, 1660–1830," *Eighteenth-Century Studies* 28 (1994–1995): 241–53, and "From 'Public' to 'Private': the Royal Funerals in England, 1500–1830," in *Mirrors of Mortality*, ed. Joachim Whaley (New York: Routledge, 1982), 61–79; John McManners's *Death and the Enlightenment: Changing Attitudes to Death Among Christians and Unbelievers in Eighteenth-Century France* (Oxford: Clarendon, 1981) is also a noteworthy study of European death rituals. In addition, see seminal works such as Clare Gittings, *Death, Burial and the Individual in Early Modern England* (London: Croom Helm, 1984); Olivia Bland, *The Royal Way of Death* (London: Constable, 1986); Julian Litten, *The English Way of Death: The Common Funeral Since 1450* (London: Robert Hale, 1991); Nigel Llewellyn, *The Art of Death: Visual Culture in the English Death Ritual c. 1500–c. 1800* (London: Reaktion, 1991); Peter C. Jupp and Glennys Howarth, eds., *The Changing Face of Death: Historical Accounts of Death and Disposal* (New York: St. Martin's Press, 1997); Mary J. Dobson, *Contours of Death and Disease in Early Modern England* (Cambridge, UK: Cambridge University Press, 1997); David Cressy, *Birth, Marriage and Death: Ritual, Religion and the Life-Cycle in Stuart England* (Oxford: Oxford University Press, 1997); Ralph Houlbrooke, *Death, Religion, and the Family in England, 1480–1750* (Oxford: Oxford University Press, 1998); Esther Schor, *Bearing the Dead: The British Culture of Mourning from the Enlightenment to Victoria* (Princeton, NJ: Princeton University Press, 1999); Ruth Richardson, *Death, Dissection and the Destitute*, 2nd ed. (Chicago: University of Chicago Press, 2000); Clare Gittings and Peter C. Jupp, eds., *Death in England: An Illustrated History* (New Brunswick, NJ: Rutgers University Press, 2000); Peter Marshall, *Beliefs and the Dead in Reformation England* (Oxford: Oxford University Press, 2002); Patricia Phillippy, *Women, Death and Literature in Post-Reformation England* (Cambridge, UK: Cambridge University Press, 2002); Lucinda M. Becker, *Death and the Early Modern Englishwoman* (Aldershot, UK: Ashgate, 2003); Peter C. Jupp, *From Dust to Ashes: Cremation and the British Way of Death* (Basingstoke, UK: Palgrave Macmillan, 2006); Allan Ingram and Leigh Wetherall Dickson, eds., *Disease and Death in Eighteenth-Century Literature and Culture: Fashioning the Unfashionable* (London: Palgrave Macmillan, 2016); Andrew Smith, *Gothic Death 1740–1914: A Literary History* (Manchester, UK: Manchester University Press, 2016); Carol Margaret Davison, ed., *The Gothic and Death* (Manchester, UK: Manchester University Press, 2017); David McAllister, *Imagining the Dead in British Literature and Culture, 1790–1848* (London: Palgrave Macmillan, 2018); Yael Shapira, *Inventing the Gothic Corpse: The Thrill of Human Remains in the Eighteenth-Century Novel* (London: Palgrave Macmillan, 2018); Kathleen M. Oliver, *Narrative Mourning: Death and Its Relics in the Eighteenth-Century British Novel* (Lewisburg, PA: Bucknell University Press, 2020).

2. Crystal B. Lake, *Artifacts: How We Think and Write About Found Objects* (Baltimore: Johns Hopkins University Press, 2020). Lake's new materialist reading demonstrates that artifacts were not witnesses of a recoverable past, and even inspired forms of suspicion. Though relics are distinct from artifacts, for my discussion, Chapter 5 which discusses "grave goods," is especially relevant.

3. Lake discusses further, "Not only were artifacts things that were understood to be capable of doing things, even when subsumed in the Enlightenment's mechanisms; they also preserved vestiges of pre-Enlightenment materialisms for an ostensibly Enlightened present" (18). See Latour, *Reassembling the Social: An Introduction to Actor–Network–Theory* (Oxford: Oxford University Press, 2005). For an extended consideration of how Latour's findings trouble interpretations of the eighteenth century, see Christina Lupton, Sean Silver, and Adam Sneed, "Introduction: Latour and Eighteenth-Century Studies." *Eighteenth-Century Theory and Interpretation* 57, 2 (2016): 165–79.

4. Peter Walmsley, "'Live to Die, Die to Live': An Introduction." *Eighteenth-Century Fiction* 21, no. 1 (2008): 1–11.

5. Elizabeth R. Napier, *Falling into Matter: Problems of Embodiment in English Fiction from Defoe to Shelley* (Toronto: University of Toronto Press, 2012), xvii. Napier's study is focused on corporeal embodiment (the living body), but her premise about fiction writers' increased need to present ideas in corporeal terms coincides with my larger argument. Punday agrees with Napier, remarking "modern storytelling, defined by the 'rise of the novel' in the eighteenth century, depends in part upon the emergence of scientific culture and resulting changes in our thinking about the body" (ix). For his study of the body, narrative theory, and the contemporary novel, see Daniel Punday, *Narrative Bodies: Toward a Corporeal Narratology* (London: Palgrave Macmillan, 2003). Juliet McMaster has a different perspective. She argues that for eighteenth-century novelists, "it was the mind as manifest in the body, rather than the body as it inhabits the mind, that was the topic of impelling interest" (xi). See Juliet McMaster, *Reading the Body in the Eighteenth-Century Novel* (New York: Palgrave Macmillan, 2004).

6. Mark Sandy, *Romanticism, Memory, and Mourning* (New York: Routledge, 2013). Sandy convincingly finds Romantic mourning in disparate poetic forms; I locate the novel as a genre that uniquely allows for complex material, scientific, and sentimental treatments.

7. For more on the movement and Hervey's *Meditations*, see Eric Parisot, *Graveyard Poetry: Religion, Aesthetics and the Mid-Eighteenth-Century Poetic Condition* (Farnham, UK: Ashgate, 2013).

8. Eric Parisot, *Graveyard Poetry*, 15.

9. James Hervey, *Meditations and Contemplations [. . .].* 26th ed. (Edinburgh: Printed for Alexander Donaldson, 1792), 36.

10. Eric Parisot, "The Work of Feeling in James Hervey's *Meditations Among the Tombs* (1746)." *Parergon* 31, no. 2 (2014): 121–35.

11. Parisot, "The Work of Feeling," 135.

12. Parisot, "The Work of Feeling," 135.

13. In addition to Smith, for a reading that stresses the difference in purpose and context between graveyard verse and Gothic fiction, see Evert Jan Van Leeuwen's "Funeral Sermons and Graveyard Poetry: The Ecstasy of Death and Bodily Resurrection." *Journal for Eighteenth-Century Studies* 32, no. 3 (2009): 353–71.

14. Parisot, *Graveyard Poetry*, 4.

15. Barbara Maria Stafford, *Body Criticism: Imaging the Unseen in Enlightenment Art and Medicine* (Cambridge, MA: MIT Press, 1991).

16. Mark Paterson, *The Senses of Touch: Haptics, Affects and Technologies* (Oxford: Berg, 2007), 2. For other relevant studies of the senses, see Michael J. Morgan, *Molyneux's Question: Vision, Touch and the Philosophy of Perception* (Cambridge, UK: Cambridge University Press, 1977); Jessica Riskin, *Science in the Age of Sensibility: The Sentimental Empiricists of the French Enlightenment* (Chicago: University of Chicago Press, 2002); *Sensible Flesh: On Touch in Early Modern Culture*, ed. Elizabeth D. Harvey (Philadelphia: University of Pennsylvania, 2003); *The Book of Touch*, ed. Constance Classen (Oxford: Berg, 2005); Mark Smith, *Sensory History: An Introduction* (Oxford: Berg, 2007); Constance Classen, *The Deepest Sense: A Cultural History of Touch* (Champaign: University of Illinois Press, 2012); Stephen Gaukroger, *The Natural and the Human: Science and the Shaping of Modernity, 1739–1841* (Oxford: Oxford University Press, 2014); Constance Classen, ed., *A Cultural History of the Senses*, 6 vols. (London: Bloomsbury Academic, 2014); Mark Paterson, *Seeing with the Hands: Blindness, Vision and Touch After Descartes* (Edinburgh: Edinburgh University Press, 2016); Carolyn Purnell, *The Sensational Past: How the Enlightenment Changed the Way We Use Our Senses* (New York: W. W. Norton, 2017).

17. Nancy Armstrong, *How Novels Think: The Limits of Individualism from 1719–1900* (New York: Columbia University Press, 2005), 29. Armstrong argues that the power of modern secular morality "draws from and authorizes those works of fiction where morality appears to emanate from the very core of an individual."

18. Thomas W. Laqueur, *The Work of the Dead: A Cultural History of Mortal Remains* (Princeton, NJ: Princeton University Press, 2015), 184. Laqueur seeks to move beyond the religious and secular debates. He pronounces, "The dead did not become secular," 186. "New" forms of enchantment emerge, "newly reconfigured rituals and practices." I agree with this cultural argument, but as my project maintains, growing secularity certainly influences how the dead are managed in fiction.

19. Roy Porter, *Flesh in the Age of Reason* (New York: W. W. Norton, 2004), 471–72. See also Lakoff and Johnson's *Philosophy in the Flesh: The Embodied Mind & Its Challenge to Western Thought* (Basic Books, 1999) and Mark Johnson's *The Meaning of the Body: Aesthetics of Human Understanding* (University of Chicago Press, 2012) for the importance of the human body in cognitive science theory, and a provocative examination of the primacy of bodily experience.

20. Jesse Molesworth, *Chance and the Eighteenth-Century Novel: Realism, Probability, Magic* (Cambridge, UK: Cambridge University Press, 2010), 2. Molesworth believes that studies that emphasize that the novel narrates a "new world of secular materialism" have focused too strongly on character and setting to the exclusion of the plot. His work seeks to "heal the rift between the magical and the secular worldviews" through its radical re-enchantment.

21. See Sarah Tindal Kareem's excellent account of enchantment and wonder in *Eighteenth-Century Fiction and the Reinvention of Wonder* (Oxford: Oxford University Press, 2014).

22. In *The Eighteenth-Century Novel and the Secularization of Ethics* (2010), Carol Stewart remarks, "Religion itself was 'secularized,' that is, it became less a system of belief or a body of doctrine and was increasingly defined as morality" (4). See Stewart's study for a helpful account of philosophical differences accorded to human agency by the various Christian factions.

23. Some examples include Ian Watt, *The Rise of the Novel: Studies in Defoe, Richardson, and Fielding* (Berkeley: University of California Press, 1957); Michael McKeon, *The Origins of the English Novel, 1600–1740* (Baltimore: Johns Hopkins University Press, 1987); Nancy Armstrong, *Desire and Domestic Fiction: A Political History of the Novel* (Oxford: Oxford University Press, 1987); Margaret Anne Doody, *The True Story of the Novel* (New Brunswick, NJ: Rutgers University Press, 1996); David Richter, *Ideology and Form in Eighteenth-Century Literature* (Lubbock, TX: Texas Tech University Press, 1999); Carol Stewart, *The Eighteenth-Century Novel and the*

Secularization of Ethics (Farnham, UK: Ashgate, 2010). Helen Thompson's *Fictional Matter: Empiricism, Corpuscles, and the Novel* (Philadelphia: University of Pennsylvania Press, 2017) convincingly argues that aspects of ontology and experience that shape empirical knowledge, through the influence of Locke, also shape the novel. Nancy Armstrong's *How Novels Think* claims that the history of the novel and the history of the modern individual are one and the same. I thoroughly agree with her assessment.

24. See Glennys Howarth, "Professionalizing the Funeral Industry in England 1700–1960" in Jupp and Howarth, eds., *The Changing Face of Death*, 120–34; Claire Gittings, *Death, Burial and the Individual in Early Modern England*; Roy Porter, "Death and Doctors in Georgian England," in *Death, Ritual, and Bereavement*, ed. Ralph Houlbrooke (London: Routledge, 1989) 77–94; and Porter's *Enlightenment: Britain and the Creation of the Modern World* (New York: Penguin, 2001).

25. Among Armstrong's works and others, Dror Wahrman's *The Making of the Modern Self: Identity and Culture in Eighteenth-Century England* (New Haven, CT: Yale University Press, 2004) is a notable study of the rise of modern individualism.

26. Elisabeth Bronfen, *Over Her Dead Body: Femininity, Death and the Aesthetic* (New York: Routledge, 1992); Janet Todd, *Gender, Art and Death* (New York: Continuum, 1993); Patricia Phillippy, *Women, Death and Literature in Post-Reformation England* (Cambridge, UK: Cambridge University Press, 2002); Maureen Daly Goggin and Beth Fowkes Tobin, eds., *Women and the Material Culture of Death* (New York: Routledge, 2013).

27. Samantha Matthews, *Poetical Remains: Poets' Graves, Bodies, and Books in the Nineteenth Century* (Oxford: Oxford University Press, 2004); Paul Westover, *Necromanticism: Traveling to Meet the Dead, 1750–1860* (London: Palgrave Macmillan, 2012); Deborah Lutz, *Relics of Death in Victorian Literature and Culture* (Cambridge, UK: Cambridge University Press, 2015).

28. Clark Lawlor, *Consumption and Literature: The Making of the Romantic Disease* (Basingstoke, UK: Palgrave Macmillan, 2006), 40.

29. Elizabeth R. Napier, *Falling into Matter*. Napier's insightful study of six novels examines how authors registered spiritual and moral reservations about the centrality of the body. Especially helpful is the chapter on *Frankenstein*, which argues for "the central role of the body in the creation of art" (xii).

CHAPTER 1

1. *The Tatler*, no. 118 (1772): 187. The Wool Act of 1699 aimed to decrease the importation of linen from overseas and required that a corpse only be dressed in a shroud made of wool.

2. Matthew Craske, *The Silent Rhetoric of the Body: A History of Monumental Sculpture and Commemorative Art in England, 1720–1770* (New Haven, CT: Yale University Press, 2008), 44. Reflecting the increased significance of the undertaker's duties, Frances Burney's *Cecilia, or, Memoirs of an Heiress* (1782), mentions the need for "the undertaker's men," and conveyance of the dead body "to town in a coffin." Demonstrating the historical interest in all things sepulchral, Richard Gough's *Sepulchral monuments in Great Britain: applied to illustrate the history of families, manners, habits and arts, at the different periods from the Norman conquest to the seventeenth century: with introductory observations*, 2 vols. (T. Payne and Son, 1787), includes fascinating details of tombs, as well as drawings of coffins, sculptures, and epitaphs.

3. William Gibson, *The Church of England, 1688–1832: Unity and Accord* (New York: Routledge, 2000), 174.

4. Rodney J. Pope, "The Eighteenth Century Church in the Wirral" (MA thesis, University of Wales, Lampeter, 1971), 69, quoted in Gibson, *The Church of England*, 174.

5. See Glennys Howarth, "Professionalising the Funeral Industry," in *The Changing Face of Death: Historical Accounts of Death and Disposal*, ed. Peter C. Jupp and Glennys Howarth (London: Palgrave Macmillan, 1997), 120–34, and Roy Porter, "Death and Doctors in Georgian England," in *Death, Ritual and Bereavement*, ed. Ralph Houlbrooke (New York: Routledge, 1989), 77–94.

6. See Craske, *Silent Rhetoric*, 40. Craske also cites the accounts of undertaker Richard Carpenter of Fleet Street (1746, 1747, 1763, 1764) held at Guildhall Library (MSS 5871), which include payments to suppliers and payments from clients.

7. Due to her debts, Countess Derby (who died March 14, 1797) was not buried in the Hamilton family vault until April 2, 1797, nearly three weeks following her death (we can assume she was embalmed in some way). *The Gentleman's Magazine* 81 (1797): 348–49.

8. Robert Hill, *The Pathway to Prayer and Pietie Contayning an Exposition of the Lord's Prayer* (London: Printed by W. S. for William Barrett, 1616), 130. Hill was minister of St. Bartholomew's London.

9. Anne Clifford, *The Diaries of Lady Anne Clifford*, ed. D. J. H. Clifford (Stroud, UK: Sutton, 1990), 74. Phillippy remarks that "Mourning garments for the aristocracy were governed by the College of Arms and their imitation by the lower social ranks influenced fashion for all classes" (254n73).

10. "Sunday 8 December 1667," *The Diary of Samuel Pepys: Daily Entries from the 17th-Century London Diary*, https://www.pepysdiary.com/diary/1667/12/08/. The first year of mourning for widows conventionally meant full black, unadorned dress for one year. Second mourning (after the first year) allowed for no cap, silk (less crape) for six to nine months, and the remaining three months of the second year plain black (with trimmings) without crape, and, in the nineteenth-century especially, jet ornaments.

11. Noted in the *Diary of the first Earl of Egmont* (London: HMSO, 1920), 448. Purple was the mourning color worn by royalty only. For more on mourning dress, see Lou Taylor, *Mourning Dress: A Costume and Social History* (London: Allen and Unwin, 1983), as well as Anne Buck's *Dress in Eighteenth-Century England* (Teaneck, NJ: Holmes & Meir, 1979).

12. Daniel Defoe, *Moll Flanders* (London: Penguin, 1989), 318.

13. Frances Sheridan, *The Memoirs of Miss Sidney Bidulph* (Ontario: Broadview, 2011).

14. Samuel Richardson, *Pamela; or, Virtue Rewarded* (Oxford: Oxford University Press), 12.

15. This is the first in a series of twelve paintings by Highmore illustrating scenes from Richardson's best-selling novel. For a detailed description of eighteenth-century mourning dress, see Taylor, *Mourning Dress*.

16. Taylor, *Mourning Dress*, 78. See also Phillis Cunnington and Catherine Lucas, *Costumes for Births, Marriages & Deaths* (London: A. and C. Black, 1972).

17. Lorraine Fletcher, "Introduction," in *Emmeline, The Orphan of the Castle* by Charlotte Smith, (Ontario: Broadview, 2003), 9–35; 162n1.

18. As portraits and mourning jewelry from the period reveal, mourning dress could be white, but more often women wore black.

19. Quoted in Richardson's *Early Works: 'Aesop's Fables', 'Letters Written to and for Particular Friends' and Other Works*, ed. Alexander Pettit (Cambridge, UK: Cambridge University Press, 2012), 523. See Peter Walmsley's excellent chapter "Death and Mourning Culture" in *Samuel Richardson in Context*, ed. Peter Sabor and Betty A. Schellenberg (Cambridge, UK: Cambridge University Press, 2017), 264–71.

20. Craske, *Silent Rhetoric*, 39. For example, Craske cites an article in the October 24, 1754, issue of the *Connoisseur* that describes a cheesemonger's interment with his "ancestors" as a parody of the laying to rest of a landed gentleman.

21. Quoted in Craske, *Silent Rhetoric*, 43. For an exceptional analysis of *Tristram Shandy*'s black page, literary epitaphs, and mourning stationery see Helen Williams, "'Alas, poor YORICK!': Sterne's Iconography of Mourning," *Eighteenth-Century Fiction* 28, no. 2 (Winter 2015): 313–44.

22. Allan Ingram and Leigh Wetherall Dickson, eds., *Disease and Death in Eighteenth-Century Literature and Culture Fashioning the Unfashionable* (London: Palgrave Macmillian, 2018), 2.

23. Eliza Haywood, *The History of Miss Betsy Thoughtless* (Oxford: Oxford University Press, 1997), 307.

24. Sarah Fielding, *The Adventures of David Simple* (Lexington: University Press of Kentucky, 1998), 36.

25. Frances Burney, *Evelina, or the History of a Young Lady's Entrance into the World* (Oxford: Oxford University Press, 2002), 41.

26. Mary Robinson, *The Widow, Or a Picture of Modern Times: A Novel, in a Series of Letters, in Two Volumes* (London, 1794). Quoted in Karen Bloom Gevirtz's *Life After Death: Widows and the English Novel, Defoe to Austen* (Newark: University of Delaware Press, 2005), 43. See Gevirtz for an assessment of fictional widows in the period. Her discussion of Sterne's Widow Wadman is especially insightful.

27. Margaret Anne Doody, *Frances Burney: The Life in the Works* (New Brunswick, NJ: Rutgers University Press, 1988), 381. In her letter to Mrs. Lock of January 9, 1800, Burney writes, "She was the soul of my soul!" (455). Incidentally, Madame d'Arblay (Frances Burney) dies on the same date in 1840, "the anniversary of that day she long consecrated to prayer, and to the memory of her beloved sister Susanna." *The diary and letters of Frances Burney, Madame d'Arblay*, ed. Susan Coolidge (Boston: Roberts Brothers, 1880), 2:540.

28. Mary Wortley Montagu, *Letters written during her travels in Europe, Asia et Africa*, 25.

29. Pierre Jacques Brillon, *Dictionnaire des arrêts, ou jurisprudence universelle des* . . . (Paris, 1728), 2:609–12.

30. Catherine Ingrassia, *Authorship, Commerce, and Gender in Early Eighteenth-Century England: A Culture of Paper Credit* (Cambridge, UK: Cambridge University Press, 1998), 25.

31. Maxine Berg, *Luxury and Pleasure in Eighteenth-Century Britain* (Oxford: Oxford University Press, 2005), 226.

32. Some of this section is adapted from my previous work. Jolene Zigarovich, "Introduction," in *Sex and Death in Eighteenth-Century Literature* (New York: Routledge, 2013), 1–27.

33. Tragedy becomes increasingly topical and politically suggestive in the last decades of the seventeenth century. After 1675, serious drama displays a noticeably darker mood, which would deepen as the decade drew toward an end. See Jean Marsden, "Tragedy and Varieties of Serious Drama," in *A Companion to Restoration Drama*, ed. Susan J. Owen (Oxford: Blackwell, 2002), 228–42.

34. For more on the intersection of sexuality and death, see Zigarovich, "Introduction," in *Sex and Death in Eighteenth-Century Literature*, 1–27.

35. Paula R. Backscheider, *Elizabeth Singer Rowe and the Development of the English Novel* (Baltimore, MD: Johns Hopkins University Press, 2013), 59.

36. See Jolene Zigarovich, "Courting Death: Necrophilia in Samuel Richardson's *Clarissa*," in *Sex and Death in Eighteenth-Century Literature*, 76–102.

37. Clark Lawlor, *Consumption and Literature: The Making of the Romantic Disease* (London: Palgrave Macmillan, 2006), 72. Lawlor argues that historically consumption is the disease of the good death, and that poet John Donne's death from consumption in 1631 was a model for Richardson's depiction of Clarissa's illness and death. While I do not read her as necessarily suf-

fering from consumption, Lawlor's consumptive interpretation of Clarissa's body is provocative. See 58–69.

38. Ian Watt, *The Rise of the Novel: Studies in Defoe, Richardson and Fielding* (Berkeley: University of Berkeley Press, 1957), 217.

39. Clarissa's wedding shroud signals her marriage to death. In Howe's *Letters Moral and Entertaining*, a "bad death" is eroticized as Amanda orders a new, fashionable wardrobe from France for her dying (*Letters*, 2:35–36). See Backscheider, Elizabeth Singer Rowe, for an insightful discussion of Rowe's work.

40. Garrett Stewart, *Death Sentences: Styles of Dying in British Fiction*. (Cambridge, MA: Harvard University Press, 1984), 25. See also Allan Wendt's "Clarissa's Coffin," *Philological Quarterly* 39 (January 1960): 481–95, for an interesting discussion of Clarissa's symbolic coffin and orthodox morality.

41. Roseann Runte, "Dying Words: The Vocabulary of Death in Three Eighteenth-Century English and French Novels," *Canadian Review of Comparative Literature* 6 (1979): 360.

42. See Kelly McGuire's evocative reading of the origins of the term "suicide," as well as the role of *Clarissa* in perpetuating the term, in *Dying to Be English: Suicide Narratives and National Identity, 1721–1814* (New York: Routledge, 2012).

43. Terry Castle, *The Female Thermometer: Eighteenth-Century Culture and the Invention of the Uncanny* (Oxford: Oxford University Press, 1995), 130.

44. Currently there is no book-length study that encompasses solely British conceptions of death and dying in the eighteenth century. For a closer understanding of changing attitudes toward death in the eighteenth century, see Paul S. Fritz's "The Undertaking Trade in England: Its Origins and Early Development, 1660–1830," *Eighteenth-Century Studies* 28 (1994–1995): 241–53, and "From 'Public' to 'Private': the Royal Funerals in England, 1500–1830" in *Mirrors of Mortality*, ed. Joachim Whaley (New York: St. Martin's, 1982), 61–79; Clare Gittings, *Death, Burial and the Individual in Early Modern England* (London: Croom Helm, 1984); Olivia Bland, *The Royal Way of Death* (London: Constable, 1986); Julian Litten, *The English Way of Death* (London: Robert Hale, 1991); Nigel Llewellyn, *The Art of Death: Visual Culture in the English Death Ritual c.1500–c.1800* (London: Reaktion, 1991); Peter C. Jupp and Glennys Howarth, eds., *The Changing Face of Death*; David Cressy, *Birth, Marriage and Death: Ritual, Religion and the Life-Cycle in Stuart England* (Oxford: Oxford University Press, 1997); Mary J. Dobson, *Contours of Death and Disease in Early Modern England* (Cambridge, UK: Cambridge University Press, 1997); Ralph Houlbrooke, *Death, Religion, and the Family in England, 1480–1750* (Oxford: Oxford University Press, 1998); Esther Schor, *Bearing the Dead: The British Culture of Mourning from the Enlightenment to Victoria* (Princeton, NJ: Princeton University Press, 1999); Ruth Richardson, *Death, Dissection and the Destitute*, 2nd ed. (Chicago: University of Chicago Press, 2000); Peter C. Jupp and Clare Gittings, eds., *Death in England: An Illustrated History* (New Brunswick, NJ: Rutgers University Press, 2000); Nancy Isenberg and Andrew Burstein, eds., *Mortal Remains: Death in Early America* (Philadelphia: University of Pennsylvania Press, 2002); Peter Marshall, *Beliefs and the Dead in Reformation England* (Oxford: Oxford University Press, 2002); Patricia Phillippy, *Women, Death and Literature in Post-Reformation England* (Cambridge, UK: Cambridge University Press, 2002); Peter C. Jupp, *From Dust to Ashes: Cremation and the British Way of Death* (London: Palgrave Macmillan, 2006). John McManners's *Death in the Enlightenment: Changing Attitudes to Death Among Christians and Unbelievers in Eighteenth-Century France* (Oxford: Oxford University Press, 1981) is a noteworthy study of European death rituals, as is the collection *Dying, Death, Burial and Commemoration in Reformation Europe*, ed. Elizabeth C. Tingle and Jonathan Willis (New York: Routledge, 2016).

45. Philippe Ariès, *The Hour of Our Death*, trans. Helen Weaver (New York: Knopf, 1981), 387. This, along with his *Western Attitudes Toward Death*, trans. Patricia Ranum (Baltimore: Johns Hopkins University Press, 1974), charts these cultural changes, albeit in sweeping terms. Jonathan Dollimore's *Death, Desire and Loss in Western Culture* (New York: Routledge, 1998) does not significantly engage the eighteenth century.

46. Michel Foucault, *The Birth of the Clinic: An Archaeology of Medical Perception* (New York: Vintage, 1994), 155.

47. Elisabeth Bronfen, *Over Her Dead Body: Death, Femininity, and the Aesthetic* (New York: Routledge, 1992), 86. See Bronfen for a larger discussion of the intersection of beauty, morbidity, and the feminine that pervades Western culture.

48. "The Burials now among the Reformed in England, are in a manner prophane," wrote N. Strange in the 1640s, "in many places the dead being throwne into the ground like dogs, and not a word said." "To the Reader," in Benjamin Carier, *A Missive to His Majesty of Great Britain . . . By Doctor Carier, Conteining the Motives of his Conversion to Catholike Religion* (Paris, 1649), n.p.

49. David Stannard, *The Puritan Way of Death: A Study in Religion, Culture, and Social Change* (Oxford: Oxford University Press, 1977), 155.

50. Peter Marshall, ed., "'The Map of God's Word': Geographies of the Afterlife in Tudor and Early Stuart England," in *The Place of the Dead: Death and Remembrance in Late Medieval and Early Modern Europe* (Cambridge, UK: Cambridge University Press, 2000), 130.

51. Voltaire, *Philosophical Letters, or Letter Regarding the English Nation*, trans. Prudence L. Steiner, ed. John Leigh (Indianapolis: Hackett, 2007), 5.

52. Eric Parisot, *Graveyard Poetry: Religion, Aesthetics and the Mid-Eighteenth-Century Poetic Condition* (Farnham, UK: Ashgate, 2013), 19.

53. John Trenchard, *A Collection of Scarce and Valuable Tracts, on the Most . . .* (London, 1748), 3:208.

54. McManners, *Death and the Enlightenment*, 202.

55. Daniel Turner, *Preparation for death explained and enforced. A sermon, occasioned by the death of Mrs. Elizabeth Tomkins . . . preached at the time of her interment* (Abingdon: printed by J. Stacey, 1769), 12. See also D. P. Walker, *The Decline of Hell: Seventeenth-Century Discussions of Eternal Torment* (University of Chicago Press, 1964) and Philip C. Almond, *Heaven and Hell in Enlightenment England* (Cambridge, UK: Cambridge University Press, 1994).

56. Ralph Houlbrooke, "The Age of Decency: 1600–1760," in *Death in England: An Illustrated History*, ed. Peter C. Jupp and Clare Gittings (New Brunswick, NJ: Rutgers University Press, 2000), 180.

57. Julie Rugg, "From Reason to Regulation: 1760–1850," in *Death in England: An Illustrated History*, ed. Peter C. Jupp and Clare Gittings (New Brunswick, NJ: Rutgers University Press, 2000), 204.

58. David Hempton notes this in *The Church in the Long Eighteenth Century: The I. B. Tauris History of the Christian Church* (London: I. B. Tauris, 2011), 158.

59. Elizabeth Rowe, *Friendship in Death: In Twenty Letters from the Dead to the Living* (1728), 22. This was published with Part I of *Letters Moral and Entertaining*, followed by Part II in 1731 and Part III in 1732. Several editions appeared throughout the century and as late as 1818. In her novellas, death must capitulate to those achieving true love. Based on classical dialogues of and from the dead, Rowe's *Friendship* as well as numerous others in the period speak to their popularity, such as Tom Brown's *Letters from the Death to the Living* (1702–1703), Matthew Prior's *Dialogues of the Dead* (1721), Daniel Defoe's *An Essay on the History and Reality of Appari-*

tions (1727), and George Lyttleton's *Dialogues of the Dead* (1760). Hester Thrale Piozzi and Anna Laetitia Barbauld both published dialogues with the dead near the end of the century. Backscheider notes that Frederick Keener lists seventy-eight dialogues published between 1641 and 1800.

60. Andrew Spicer, "'Defyle not Christ's Kirk with your Carrion': Burial and the Development of Burial Aisles in Scotland," in *The Place of the Dead: Death and Remembrance in Late Medieval and Early Modern Europe*. (Cambridge, UK: Cambridge University Press, 2000), 150.

61. "The Sermons of John Wesley: Sermon 73, Of Hell," Wesley Center Online, accessed July 10, 2021, http://wesley.nnu.edu/john-wesley/the-sermons-of-john-wesley-1872-edition/sermon-73-of-hell/

62. Ralph Houlbrooke, *Death, Religion, and the Family*, 34.

63. Patricia A. Whiting, "Women's Magazines and the Representation of Death in Eighteenth-Century England" (PhD diss., University of Ottawa, 1999), 88.

64. Stannard, *Puritan Way of Death*, 150.

65. See Michel Foucault, *The History of Sexuality*, vol. 1, *An Introduction* (New York: Vintage, 1990), 138–39.

66. Cited in Stannard's *The Puritan Way of Death*, 147. Stannard continues, "It was becoming the norm, the accepted norm, for the godly to die . . . 'in Raptures of holy Joy,'" 150.

67. Jeremy Taylor, *The Rule and Exercises of Holy Dying . . .*, 24th ed. (London, 1727), 38.

68. Adam Smith, *Theory of Moral Sentiments* (1759), from "Chapter I: Of Sympathy, Section I: Of the Sense of Propriety," https://oll.libertyfund.org/title/smith-the-theory-of-moral-sentiments-and-on-the-origins-of-languages-stewart-ed.

69. Jeremy Gregory, "Religion in the Age of Enlightenment: Putting John Wesley in Context," in *Religion in the Age of Enlightenment*, ed. Brett C. McInelly (New York: AMS Press, 2010), 2:22. The seminal work of Alan D. Gilbert, Roy Porter, and Peter Gay is certainly in this camp.

70. The radical camp would include Margaret C. Jacob's *The Radical Enlightenment—Pantheists, Freemasons and Republicans* (London: George Allen and Unwin, 1981), Jonathan I. Israel's *Radical Enlightenment: Philosophy and the Making of Modernity 1650–1750* (Oxford: Oxford University Press, 2001), and Steffen Ducheyne, ed., *Reassessing the Radical Enlightenment* (New York: Routledge, 2017).

71. Phyllis Mack, "The Senses in Religion: Listening to God in the Eighteenth Century," in *A Cultural History of the Senses in the Age of Enlightenment*, ed. Anne C. Vila (London: Bloomsbury, 2014), 85.

72. Alex Schulman, *The Secular Contract: The Politics of Enlightenment* (London: Bloomsbury, 2013), 1.

73. Margaret C. Jacob, *The Secular Enlightenment* (Princeton, NJ: Princeton University Press, 2019), 1.

74. Stephen Toulmin, *Cosmopolis: The Hidden Agenda of Modernity* (Chicago: University of Chicago Press, 1990), 37.

75. Isaac Newton, *The Mathematical Principles of Natural Philosophy and His System of the World* (Principia), trans. Andrew Motte (Berkeley: University of California Press, 1966), book III.395.

76. Stewart J. Brown and Timothy Tackett, Introduction, in *The Cambridge History of Christianity*, vol. 7, *Enlightenment, Reawakening and Revolution, 1660–1815* (Cambridge, UK: Cambridge University Press, 2006), 5.

77. Jane Shaw, *Miracles in Enlightenment England* (New Haven, CT: Yale University Press, 2006). See especially 77–82, 161–77.

78. Jeremy Gregory, "Religion in the Age of Enlightenment," 53.

79. Ulrich L. Lehner, *The Catholic Enlightenment: The Forgotten History of a Global Movement* (Oxford: Oxford University Press, 2016), 3.

80. David Sorkin, *The Religious Enlightenment: Protestants, Jews, and Catholics from London to Vienna* (Princeton, NJ: Princeton University Press, 2008), 3. There are numerous religious tracts by moral philosophers in the period, including Locke's *The Reasonableness of Christianity* (1695). Rational religion is found in Matthew Tindal's *Christianity as Old as the Creation* (1730), which posits that the gospel is simply a republication of the law of nature. Advocates of natural religion and moralism believed their version of religion could be validated by its utility.

81. See Jeremy Gregory's "Religion: Faith in the Age of Reason," *Journal for Eighteenth-Century Studies* 34 no. 4 (2011): 435–43, for an excellent critical review of post-1970s work on religion in the period.

82. It is important to note that numerous *philosophes* and British thinkers (to include Shaftesbury, a student of Locke's) did not embrace empirical methods and were "anti-body," rejecting theories of knowledge gained through the senses.

83. Joseph Addison, "Pleasures of Imagination," *Spectator*, no. 411, June 21, 1712, https://www.ourcivilisation.com/smartboard/shop/fowlerjh/chap22.htm. My thanks to Daniel O'Quinn for recommending the inclusion of Addison's quotation.

84. Hannah More, *Strictures on the Modern System of Female Education* (London, 1799), 2:68.

85. Edmund Burke, *The Works of the Right Honourable Edmund Burke*, vol. 1 (London, 1887; Project Gutenberg, 2005), http://www.gutenberg.org/files/15043/15043-h/15043-h.htm.

86. Jonathan Reinarz and Leonard Schwarz, "The Senses and the Enlightenment: An Introduction," *Journal for Eighteenth-Century Studies* 35, no. 4 (2012): 465.

87. John Locke, *An Essay Concerning Human Understanding* (1689), https://www.gutenberg.org/files/10615/10615-h/10615-h.htm.

88. *An Essay Concerning Human Understanding*, II.4, "The Idea of Solidity," https://www.gutenberg.org/files/10615/10615-h/10615-h.htm.

89. Though this extends beyond my project, I wish to acknowledge that eighteenth-century discourses on the senses were deeply influenced by classical texts (Aristotle's *De Sensu*, *De Anima*, Empedocles, Democritus, Plato's *Republic*, *Theaetetus*), which would be required reading for many university students. For helpful discussions of eighteenth-century theories of the senses, see Michael J. Morgan, *Molyneux's Question: Vision, Touch and the Philosophy of Perception* (Cambridge, UK: Cambridge University Press, 1977); Michael Baxandall, *Shadows and Enlightenment* (New Haven, CT: Yale University Press 1995); M. Degenaar, *Molyneux's Problem: Three Centuries of Discussion on the Perception of Forms* (New York: Springer, 1996); Jessica Riskin, *Science in the Age of Sensibility: The Sentimental Empiricists of the French Enlightenment* (Chicago: University of Chicago Press, 2002); Shaun Gallagher, *How the Body Shapes the Mind* (Oxford: Clarendon Press, 2005); Mark Paterson, *Seeing with the Hands: Blindness, Vision and Touch After Descartes* (Edinburgh: Edinburgh University Press, 2016).

90. Though I do not have the space here to elucidate the previous debates between the mind and body, it is relevant to note that in the previous century, Montaigne (a practicing Catholic) explains that philosophers are drawn to dualism when they are uncomfortable with their own corporeal natures (Toulmin, 44). Montaigne writes about this fear of the flesh: "Philosophy is very childish, to my mind, when she gets up on her hind legs and preaches to us that is a barbarous alliance to marry the divine with the earthly, the reasonable with the unreasonable, the sever

with the indulgent, the honorable with the dishonorable; that sensual pleasure is a brutish thing unworthy of being enjoyed by the wise man" (*The Complete Works, Essays*, 1042).

91. David Hume, *A Treatise of Human Nature* (1738), https://www.gutenberg.org/files/4705/4705-h/4705-h.htm.

92. John W. Yolton, *Thinking Matter: Materialism in Eighteenth-Century Britain* (Minneapolis: University of Minnesota Press, 1984), 82.

93. Denis Diderot, *Letter on the Blind* (*Lettre sur les aveugles à l'usage de ceux qui voient*, 1749). Kate E. Tunstall, *Blindness and Enlightenment. An Essay. With a New Translation of Diderot's Letter on the Blind* (London: Continuum, 2011).

94. Edmund Burke, *A Philosophical Enquiry into the Origin of our Ideas of the Sublime and Beautiful* (1757), last modified July 5, 2021, https://www.gutenberg.org/files/15043/15043-h/15043-h.htm.

95. Mark Paterson, *Seeing with the Hands: Blindness, Vision and Touch After Descartes* (Edinburgh: Edinburgh University Press, 2016), 48.

96. George Berkeley, "An Essay Toward a New Theory of Vision," in *Philosophical Works, Including the Works on Vision*, ed. Michael R. Ayers (London: J. M. Dent, 1975).

97. Thirty-six years after the publication of Molyneux's question, English surgeon William Cheselden published an account of the result of cataract surgery on a blind boy that seemed to confirm this prediction. When the boy's sight returned, he did not know shapes and could not distinguish things regardless of shape or size. See "An Account of some Observations made by a young Gentleman, who was born blind, or lost his Sight so early, that he had no Remembrance of ever having see, and was couch'd between 13 and 14 Years of Age," *Philosophical Transactions* 402 (1728): 447–50.

98. Étienne Bonnot, Abbé de Condillac, *Treatise on Sensations* (*Traité des sensations*, 1754), Stanford Encyclopedia of Philosophy, last modified September 21, 2017, https://plato.stanford.edu/entries/condillac/. For more on the politicization of touch, see Erin Manning's *Politics of Touch: Sense, Movement, Sovereignty* (Minneapolis: University of Minnesota Press, 2006). Manning provides an insightful summary of philosophical theories of touch and the senses (including Condillac).

99. For a helpful translation and explanation, see the *Stanford Encyclopedia of Philosophy*. https://plato.stanford.edu/index.html.

100. Jean-Jacques Rousseau, *Émile* (London: J. M. Dent, 1966), 103.

101. Suzanne Clover Lindsay, "Mummies and Tombs: Turenne, Napoleon, and Death Ritual," *The Art Bulletin* 82, no. 3 (2000): 479.

102. Mark Paterson, *The Senses of Touch: Haptics, Affects and Technologies* (Oxford: Berg, 2007), 2. For other relevant discussions, see Michael J. Morgan, *Molyneux's Question. Vision, Touch and the Philosophy of Perception* (Cambridge, UK: Cambridge University Press, 1977); Jessica Riskin, *Science in the Age of Sensibility: The Sentimental Empiricists of the French Enlightenment* (Chicago: University of Chicago Press, 2002); *Sensible Flesh: On Touch in Early Modern Culture*, ed. Elizabeth D. Harvey (Philadelphia: University of Pennsylvania, 2003); *The Book of Touch*, ed. Constance Classen and David Howes (Oxford: Berg, 2005); Mark Smith, *Sensory History: An Introduction* (Oxford: Berg, 2007); Constance Classen, *The Deepest Sense: A Cultural History of Touch* (Chicago: University of Illinois Press, 2012); Stephen Gaukroger, *The Natural and the Human: Science and the Shaping of Modernity, 1739–1841* (Oxford: Oxford University Press, 2014); *A Cultural History of the Senses*, 6 vols., ed. Constance Classen (London: Bloomsbury, 2014); Mark Paterson, *Seeing with the Hands: Blindness, Vision and Touch After Descartes*

(Edinburgh: Edinburgh University Press, 2016); Carolyn Purnell, *The Sensational Past: How the Enlightenment Changed the Way We Use Our Senses* (New York: Norton, 2017).

103. Zahra Newby and Ruth E. Toulson, eds, introduction to *The Materiality of Mourning: Cross-Disciplinary Perspectives* (New York: Routledge, 2018), 1.

104. Helpful studies include John Mullan, *Sentiment and Sociability: The Language of Feeling in the Eighteenth Century* (Oxford: Clarendon Press, 1988); G. J. Barker-Benfield, *The Culture of Sensibility: Sex and Society in Eighteenth-Century Britain* (Chicago: University of Chicago Press, 1996); Anne C. Vila, *Enlightenment and Pathology: Sensibility in the Literature and Medicine of Eighteenth-Century France* (Baltimore: Johns Hopkins University Press, 1998); Michael Bell, *Sentimentalism, Ethics and the Culture of Feeling* (London: Palgrave Macmillan, 2000); Julie McMaster, *Reading the Body in the Eighteenth-Century Novel* (London: Palgrave Macmillan, 2004); Ann Jessie Van Sant, *Eighteenth-Century Sensibility and Novel: The Senses in Social Context* (Cambridge, UK: Cambridge University Press, 2004); Paul Goring, *The Rhetoric of Sensibility in Eighteenth-Century Culture* (Cambridge, UK: Cambridge University Press 2005); Ildiko Csengei, *Sympathy, Sensibility and the Literature of Feeling in the Eighteenth Century* (London: Palgrave Macmillan, 2012); Helen Deutsch and Mary Terrall, eds., *Vital Matters: Eighteenth-Century Views of Conception, Life, and Death* (Toronto: University of Toronto Press, 2012); Catherine Packham, *Eighteenth-Century Vitalism: Bodies, Culture, Politics* (London: Palgrave Macmillan, 2012); Henry Martyn Lloyd, ed., *The Discourse of Sensibility: The Knowing Body in the Enlightenment* (New York: Springer, 2013); Alex Wetmore, *Men of Feeling in Eighteenth-Century Literature: Touching Fiction* (London: Palgrave Macmillan, 2013); Amit S. Yahav, *Feeling Time: Duration, the Novel, and Eighteenth-Century Sensibility* (Philadelphia: University of Pennsylvania Press, 2018); Soile Ylivuori, *Women and Politeness in Eighteenth-Century England: Bodies, Identities, and Power* (New York: Routledge, 2018).

105. Roy Porter, *English Society in the Eighteenth Century* (London: Penguin, 1982), 323.

106. Benjamin Barker-Benfield, *The Culture of Sensibility: Sex and Society in Eighteenth-Century Britain* (University of Chicago Press, 1992).

107. Anne C. Vila, "Introduction: Powers, Pleasures, and the Perils of the Senses in the Enlightenment Era," in *A Cultural History of the Senses in the Age of Enlightenment*, ed. Anne C. Vila (London: Bloomsbury, 2014), 7.

108. For more on the connection between the senses and literature, see the references noted above, as well as Kevin L. Cope and Robert C. Leitz, eds., *Sensational Centuries: Essays on the Enhancement of Sense Experience in the Seventeenth, Eighteenth, and Nineteenth Centuries* (New York, AMS Press, 2013); Rowan Rose Bryson, "The Senses in Literature: Pleasures of Imagining in Poetry and Prose," in *A Cultural History of the Senses in the Age of Enlightenment* (Bloomsbury, 2014), 155–78; Mary Helen McMurran and Alison Conway, eds., *Mind, Body, Motion, Matter Eighteenth-Century British and French Literary* (Toronto: University of Toronto Press, 2016); Emily C. Friedman, *Reading Smell in Eighteenth-Century Fiction* (Lewisburg, PA: Bucknell University Press, 2016). Helen Thompson, *Fictional Matter: Empiricism, Corpuscles, and the Novel* (Philadelphia: University of Pennsylvania Press, 2017), is also an engaging account of empiricism's influence on the emerging novel.

109. Bryson, "The Senses in Literature," 168.

110. David Hartley, *Observations on Man, his Frame, his Duty, and his Expectations* (1749). Incidentally, it was published in two parts by Samuel Richardson. Hartley was an empiricist and theorized that external stimuli produced vibrations in the brain.

111. John Mullan, *Sentiment and Sociability: The Language of Feeling in the Eighteenth Century* (Oxford: Clarendon Press, 2002), 201. As Mullan and others note, sentimentalism fo-

cuses on the text's treatment of the body. For these novels, bodies are fully legible and revelatory.

112. Craske, *Silent Rhetoric*, 313.

113. Craske, *Silent Rhetoric*, 313.

114. Henry Mackenzie, *The Man of Feeling* (New York: W. W. Norton, 1951), 92.

115. Lawlor, *Consumption and Literature*, 49.

116. For further discussion, see Lawlor, *Consumption and Literature*, 50–58.

117. See McGuire, *Dying to Be English*, for an evocative reading of gender and suicide in British literature and culture.

118. For a helpful discussion, see Bryson, "The Senses in Literature." On the "Werther-effect" in Britain, see McGuire's chapter, "The Pathology of Sentiment: Politics, Sacrifice and Wertherism in the English Novel of Sensibility," in *Dying to Be English*, 115–49.

CHAPTER 2

1. An early version of this chapter was previously published in *Eighteenth-Century Life* 33, no. 3 (2009): 65–104. It is reprinted by kind permission of Duke University Press. Samuel Richardson, *Clarissa, or The History of a Young Lady* [1751], 3rd ed., 8 vols. (London: Printed for S. Richardson and sold by John Osborn, 1747–1748), 8:45, letter 17, Mr. Lovelace to Mr. Belford. All subsequent quotes are from this edition.

2. Elisabeth Bronfen, *Over Her Dead Body: Death, Femininity, and the Aesthetic* (New York: Routledge, 1992), 95. Margaret Anne Doody, in *A Natural Passion: A Study of the Novels of Samuel Richardson* (Oxford: Clarendon, 1974), likens this scene to Nat Lee's *Caesar Borgia* (1680): "Like Lee's Borgia, he [Lovelace] wishes to have embalmed the body of the woman he has once taken against her will, and whose death he has caused, a grim parallel to the possession of the body but not the person in the act of rape" (121).

3. Richardson's religious views are certainly relevant to discussions of death and burial practices in his novel—the author himself described *Clarissa* to Lady Bradshaigh as a "Religious Novel" (October 26, 1748), and Clarissa's excruciating, prolonged death exemplifies an Anglican "good death"—yet I agree with E. Derek Taylor that in his novels, Richardson is "reticent and ecumenical in matters of religion," and that there exists the potential for several other theological influences (Puritan, mystic). Not wishing to enter this lengthy scholarly debate, my discussion of the novel seeks to trace the growing sentimental attitude toward the corpse, as well as the influence of empirical inquiry on burial practices. For helpful discussions of Richardson's religious beliefs and his novels, see E. Derek Taylor, *Reason and Religion in* Clarissa*: Samuel Richardson and 'the Famous Mr. Norris, of Bemerton'* (New York: Routledge, 2017), as well as his chapter on religion in *Samuel Richardson in Context* (Cambridge, UK: Cambridge University Press, 2017), 302. For a broader discussion of religion and death in the period, see Chapter 1 in this volume.

4. I have argued this elsewhere, in particular "Courting Death: Necrophilia in Samuel Richardson's *Clarissa,*" *Studies in the Novel* 32, no. 2 (2000): 112–28, revised and reprinted in *Sex and Death in Eighteenth-Century Literature* (New York: Routledge, 2013), 76–102. See also the introduction to this volume.

5. Though dated and prone to some embellishment, Jessie Dobson's "Some Eighteenth-Century Experiments in Embalming," *Journal of the History of Medicine* 8 (1953): 431–41, is a fascinating glimpse into eighteenth-century British embalming practices. An essay that includes a

discussion of the cultural view of embalming in the late seventeenth and early eighteenth centuries is Robert A. Aubin's "Behind Steele's Satire on Undertakers," *PMLA* 64 (1949): 1009–26. Patricia Phillippy's *Women, Death, and Literature in Post-Reformation England* (Cambridge, UK: Cambridge University Press, 2002) includes a discussion of the embalming of early royals.

6. Terry Castle, *The Female Thermometer: Eighteenth-Century Culture and the Invention of the Uncanny* (Oxford: Oxford University, 1995), 130. It must be noted that unfortunately, Castle does not provide specific evidence to support this claim. Though I would amend her statement, for the practice was clearly not "common," I would argue that the medical discourse and experimentation with embalming, mainly for the benefit of anatomical purposes, was indeed prevalent. Castle's description must be corrected as well: Bentham's *head* was embalmed, not his corpse.

7. Aubin, "Behind Steele's Satire," 1012. Aubin's essay provides a brief history of seventeenth- and eighteenth-century embalming practices and includes a helpful summary of social attitudes toward embalming.

8. Stephen Porter, "Death and Burial in Seventeenth-Century Oxford," *Oxfordshire Local History* 1 (1980): 2–7.

9. Stephen Porter, "Death and Burial in a London Parish: St Mary Woolnoth 1653–99," *London Journal* 8, no. 1 (1982): 76–80.

10. Leslie Clarkson, *Death, Disease, and Famine in Pre-Industrial England* (New York: St. Martin's Press, 1976), 153. Fifty guineas would amount to close to £15,000 today.

11. Previous to my findings, some death historians minimized the occurrence and interest in embalming. In *The Disposal of the Dead* (New York: Philosophical Library, 1953), C. J. Polson states, "Embalming was never common in Britain" (217). In *The English Way of Death* (London: Robert Hale, 1991), Litten claims that "embalming was rarely practiced during the eighteenth century," yet he has unearthed many instances of embalming in England, which he includes in a chapter devoted to its history. In "The Undertaking Trade," Paul S. Fritz agrees: "Embalming was rare and was employed only when there was a long lapse of time between the individual's death and the funeral ceremony" (242).

12. Elizabeth Hallam and Jenny Hockey, *Death, Memory and Material Culture* (Oxford: Berg, 2001), 132.

13. See Brian Parsons, *The Evolution of the British Funeral Industry in the 20th Century: From Undertaker to Funeral Director* (Bingley, UK: Emerald, 2018), 53.

14. See the website for the British Institute of Embalmers (http://www.bioe.co.uk/) for a brief history of embalming methods and practices.

15. Robert Latham and William Matthews, eds., *The Diary of Samuel Pepys*, 11 vols. (Berkeley: University of California Press, 1976), 9:457. Samuel Pepys, in his entry for February 23, 1669, recorded that on his thirty-sixth birthday, he kissed the embalmed body of Queen Katherine, dead since 1437. "I now took them [his wife and girl servants] to Westminster Abbey and there did show them all the tombs very finely, having one with us alone . . . and here we did see, by perticular [*sic*] favour, the body of Queen Katherine of Valois, and had her upper part of her body in my hands. And I did kiss her mouth, reflecting upon it that I did kiss a Queen, and that this was my birthday, 36 year old, that I did first kiss a Queen" (9:457). By 1669, Katherine de Valois's corpse was housed in a single wooden shell, in the north aisle of Henry VII's Chapel, further testament to the quality of the embalmment. The body was eventually buried in 1778, and a century later, Dean Stanley removed her remains for permanent burial under the altar in Henry V's chantry. Her painted wooden funeral effigy is on display in the abbey's museum, but the effigy used at Henry's funeral has not survived. See http://www.westminster-abbey.org/.

16. *The London Gazette*, January 5, 1681, 2.

17. According to Charles Bradford, in *Heart Burial* (London: Allen and Unwin, 1933), Queen Anne's heart was "encased in silver and covered with purple velvet" (220), then placed on her coffin in Westminster Abbey. (Bradford cites Scott's *Life of Swift* as his source.) Dr. Richard Mead primarily worked with infectious diseases.

18. Princess Charlotte was embalmed and her viscera placed in an urn in 1817. Queen Victoria's predecessor, William IV, was embalmed in 1837; he was the last British monarch to be embalmed. Queen Victoria criticized the invasive treatment of the royal corpses. Her opinion was also influenced by the awareness of the deteriorating effects of embalming fluids on cemetery soil and water supplies. One of the latest cases in the British Isles of evisceration and embalmment at death is that of Alexander Douglas, Duke of Hamilton (d. 1852), who was buried with pomp in an Egyptian sarcophagus of basalt in a classical mausoleum he had erected near Hamilton Palace. The writer in *The Gentleman's Magazine* vol. 193 (July–December 1852): 424–25, describing the funeral, remarks, "His body underwent the process of embalming, a process of which even Royalty has of late years judged proper to decline" (425).

19. Perhaps the most notorious exhumations from the period occurred during the destruction of tombs and mausoleums of former kings in the new French Republic in 1793, including that of Henry IV (d. 1610) at St. Denis. A death mask was made of the royal when he was exhumed. The government-ordered exhumation of Napoleon's corpse on October 15, 1840, though brief, was well-publicized and profoundly emotional due to his well-preserved cadaver (perhaps arsenic, an embalming agent, assisted this). See Suzanne Glover Lindsay, *Funerary Arts and Tomb Cult: Living with the Dead in France, 1750–1870* (Farnham, UK: Ashgate, 2012).

20. Crystal B. Lake, *Artifacts: How We Think and Write About Found Objects* (Baltimore: Johns Hopkins University Press, 2020), 166.

21. Charles Collingnon, "Some account of a body lately found in uncommon preservation, under the ruins of the abbey, at St. Edmund's-Bury, Suffolk; with some reflections upon the subject," *Philosophical Transactions* 62 (1772): 465–68.

22. See Litten, *The English Way of Death*, 53–54.

23. Margaret Cox, *Life and Death in Spitalfields, 1700 to 1850* (York: Council for British Archaeology, 1996), 13.

24. For example, in his letter to Horace Walpole dated October 1, 1774, Horace Mann describes the dissection and embalming of the late Pope Pius VI, who was reputably poisoned. "Never was Poison more manifestly proved than in the case of the late Pope; for, though the dissection of his Body was made the same day he died the putrefaction of it was so great that the Surgeons had great difficulty to perform that operation. As soon as they touched his head, all the hair and all his teeth fell out. All the bones of his body wasted away; and the flesh, upon the least touch of the knife, did not divide, but came away in pieces. They were forced to embalm the body twice, on order to transport it to the Vatican; but his face was so disfigured that it was necessary to cover it with a mask of wax. At the dissection, his bowels were put into a large, strong, earthen pan, but the fermentation of them was so great that it broke it." See *'Mann' and Manners at the Court of Florence, 1740–1786: Founded on the Letters to Horace Mann to Horace Walpole* (1876), 2:269.

25. See Henrietta Ponsonby, *Lady Bessborough and Her Family Circle* (London: J. Murray, 1940), 23.

26. Thomas Doolittle (1632[?]–1707), *A Sermon on eyeing of eternity, so that it may have its due influence upon us in all we do*, 2nd ed. (London, 1755), 74.

27. John Bowden (d. 1750), *The grave, the house appointed for all living. A sermon preach'd at the funeral of Mr. James Blunt, of Frome on Thursday the 19th of August, 1749* (Bristol, UK: Printed for Thomas Cadell, 1749), 4.

28. Books on how to die well peaked in the seventeenth century, but the eighteenth century saw a steady stream of Drelincourt translations and publications, such as John Kettlewell's *Death made Comfortable: or the Way to Die Well* (1702), John Hayward's *The Horrors and Terrors of the Hour of Death and Day of Judgment* (1707), James Hervey's *Meditations among the Tombs* (1746), and George Wright's *Pleasing Melancholy, or a Walk among the Tombs in a Country Church Yard* (1793).

29. David Cressy, *Birth, Marriage and Death: Ritual, Religion and the Life-Cycle in Stuart England* (Oxford: Oxford University Press, 997); Ralph Houlbrooke, *Religion, and the Family in England, 1480–1750* (Oxford: Oxford University Press, 1998).

30. As Peter C. Jupp explains in *From Dust to Ashes: Cremation and the British Way of Death*, "Britain was the first modern European country to popularise cremation and for this it has a unique place in the sociology of death. Cremation was first promoted in 1874 and enshrined in the Cremation Act of 1902" (ix). Jupp cites burial as a public health issue, laws providing access to burial and cremation, the extravagance of funeral expenditure, rapid urbanization of housing and the need for land space, and the rise of secularization as the main reasons for the growing popularity of cremation in British burial practices. In 1963, Pope Paul VI lifted the ban on cremation and in 1966 allowed Catholic priests to officiate at cremation ceremonies.

31. Ralph Brownrig, *Sixty Five Sermons*, 2 vols. (London: Printed by Tho. Roycroft for John Martyn and are to be sold by Robert Boulter, 1674), 2:210.

32. Edward Dent, *Everlasting Blessedness. A Sermon Preached at the Funeral of William Baker.* (London: Printed for W. Marshall, 1692), 20.

33. Richard Hooker, *Of the laws of ecclesiastical politie*, 2 vols. (London: Printed and sold by Benjamin Bragg, at the Blue Ball in Avemary-Lane, 1705), 2:403; John Pearson, *An Exposition of the Creed*, 12th ed. (London: Printed by Roger Daniel for John Williams [. . .], 1741), 223. Quoted by Aubin, "Behind Steele's Satire," 1010.

34. Francis Bancroft, *A True Copy of the remarkable Last Will and Testament of Mr. Francis Bancroft, Citizen and Draper of London* (London: Printed for J. Peele, 1728), 6.

35. The *Ladies Magazine, Or, The Universal Entertainer*, August 22, 1752, 3.

36. For Douce's history, last will and testament, and the start and history of the development of his manuscript collection, see *The Douce Legacy, An Exhibition to Commemorate the 150th Anniversary of the Bequest of Francis Douce (1757–1834)* (Oxford: Bodleian Library, 1984). This passage from Douce's will is quoted in Matthew Craske, "Entombed like an Egyptian: An Eighteenth-Century Surgeon's Extravagant Mausoleum to Preserve His Mortal Remains," *Church Monuments* 15 (2000): 85.

37. This extravagant burial was clearly inspired by ancient Egyptian embalming and burial practices. Douce was, notably, one of the subscribers to Thomas Greenhill's *The Art of Embalming*, and together Douce and Greenhill were members of the Company of Surgeons and embalming practitioners themselves. (See Craske, "Entombed like an Egyptian.") For other examples of burial vaults, see Julian Litten, "Tombs Fit for Kings: Some Burial Vaults of the English Aristocracy and Landed Gentry of the Period 1650–1850," *Church Monuments* 14 (1999): 104–28.

38. For a full discussion of Egypt and Europe, and the history of the Egyptianizing of European architecture, see James Stevens Curl's *Egyptomania: The Egyptian Revival, A Recurring Theme in the History of Taste* (Manchester, UK: Manchester University Press, 1994).

39. Greek and Roman styles were popular among the aristocratic, and the funeral monuments of the wealthy reflected this. The McLennan Monument (c. 1893) erected in Anfield Cemetery in the form of an Egyptian pylon is a notable later example that includes engraved Egyptian-style decoration. Yet despite the increasing interest in ancient Egypt during the nine-

teenth century, compared with other funerary styles, Egyptian monuments and commemorative statues in cemeteries were relatively rare, exotic, and costly. The continuance of this style into the early twentieth century, and the survival of these monuments into the twenty-first, signals an ongoing curiosity surrounding memorials and the Egyptian style in Britain. For an examination of eighteenth- and nineteenth-century English pyramids, see Jolene Zigarovich, "Egyptomania, English Pyramids and the Quest for Immortality," in *Ancient Egypt in the Modern Imagination*, ed. Eleanor Dobson and Nichola Tonks (London: Bloomsbury, 2020), 105–16.

40. *The London Gazette* (1682).

41. This advertisement is placed in *The London Gazette* for several months, and no subsequent information about a family claiming Francis Merrydith is given.

42. Anonymous, *The Death, Dissection, Will, and Funeral Procession of Mrs. Regency* (London: printed at the [Lo]gographic Press, and sold by J. Walter; C. Stalker; and W. Richardson, 1789), 7.

43. For a broader discussion of bowels and abdomens in the culture, and how bowels played a crucial part in eighteenth-century emotions and perceptions of the self, see *Bellies, Bowels and Entrails in the Eighteenth Century*, ed. Rebecca Anne Barr, Sylvie Kleiman-Lafon, and Sophie Vasset (Manchester, UK: Manchester University Press, 2018). In this same collection, see Jacques Gélis, "The Saints of the Entrails and the Bowels of the Earth," 311–31, for a history of Catholic intestinal saints (entephorous saints). While the attachment to local and regional intestinal saints was popular in France during this period (disorders of the gut provoked extreme anxiety), Gélis remarks that the French Catholic Church did not encourage the worship of these patron saints (314). It must also be noted that the Catholic Church condemned the exploration of the human body and considered dissection sacrilegious.

44. See Nigel Llewellyn, *The Art of Death: Visual Culture in the English Death Ritual, c. 1500–c.1800* (London: Reaktion Books, 1991), 99. Both Sir Thomas Brown, in *Hydriotaphia: Urne-Buriall, or a Discourse of The Sepulchrall Urns Lately Found in Norfolk* (London, 1658), and Alison Kelly, in "The Storied Urn," from *Decorative Wedgwood* (London: Country Life, 1965), detail this practice. The memorial (or cinerary urn) to Henry Earle was formerly held at the Wedgwood Museum Trust and is now part of the V&A Wedgwood Collection. The urn is black encaustic painted in white and green and is inscribed "Memoriae S Henrici Earle Vixit Annos LXI Mort: Ian Die XXXI AD MDCCLXXIII." Height 14 ¾". Item 4926. During this period, Wedgwood also produced decorative Egyptian-style canopic jars as well as plaster death masks.

45. Matthew Holbeche Bloxam, *A glimpse at the monumental architecture and sculpture of Great Britain, from the earliest period to the eighteenth century* (W. Pickering, 1824), 63.

46. William Salmon, *Synopsis Medicinæ, or a Compendium of Astrological, Galenical, and Chymical Physick* (London: W. Godbid, 1671), 393.

47. Letter dated December 2, 1805, quoted in *Notes and Queries,* November 17, 1849, on the publication of Pettigrew's edition of Nelson's letters. 3:36–37. It is perhaps ironic that later, Nelson's own corpse is pickled in brandy by Dr. Beatty after he dies on October 21, 1805. His body is later transferred to a lead coffin in order to make the lengthy journey to England intact for an elaborate funeral. See Roger Knight, *The Pursuit of Victory: The Life and Achievement of Horatio Nelson* (New York: Basic Books, 2005).

48. *La Belle assemblée, or, Bell's court and fashionable magazine*, August 1, 1817, 304.

49. For fascinating discussions of the gut's significance in the period, see Barr et al, eds., *Bellies, Bowels and Entrails in the Eighteenth Century.*

50. Shirley Bury, *An Introduction to Sentimental Jewellery* (London: HMSO, 1985), 6.

51. See Bertram S. Puckle, *Funeral Customs: Their Origin and Development* (London: T. Werner Laurie, 1926), and Bradford, *Heart Burial*, for detailed histories of heart burial.

52. See R. C. Finucane's "Sacred Corpse, Profane Carrion: Social Ideals and Death Rituals in the Later Middle Ages," in *Mirrors of Mortality: Studies in the Social History of Death*, ed. Joachim Whaley (New York: St. Martin's, 1981), 40–60, for a discussion of attempts to embalm in the Middle Ages.

53. See Bradford's oddly fascinating *Heart Burial* for a detailed account of notable British figures who had their hearts embalmed.

54. Bradford, *Heart Burial*, 213. The remains were placed in the Stuart vault beneath the tomb of Mary, Queen of Scots. The Duke of Gloucester's coffin lies on top of that of Elizabeth of Bohemia.

55. St. Paul's Church notes that the date of death is actually wrong. Sir Crisp died February 26, 1666. See "St. Paul's Church, Hammersmith," http://www.sph.org.

56. See *Book of Days*, ed. R. Chambers (London: W. and R. Chambers, 1869), 2:418, for a detailed account of the ceremony. Quoted in Bradford, *Heart Burial*, 235. See http://www.thebookofdays.com/months/oct/5.htm.

57. The Twickenham Museum website is http://twickenham-museum.org.

58. My thanks to Robert Folkenflik, who pointed me to this eighteenth-century depiction of heart burial.

59. The veracity of specimens supposedly taken during the autopsy (including a claim that Napoleon's penis was removed) has been disputed. See Andrew Roberts, *Napoleon: A Life* (London: Penguin, 2015).

60. Incidentally, Mary's mother, Mary Wollstonecraft, was offended by the primitive practice of embalming. In *Letters Written During a Short Residence in Sweden, Norway, and Denmark* (1796), she writes, "A desire of preserving the body seems to have prevailed in most countries of the world, futile as it is to term it a preservation, when the noblest parts are immediately sacrificed merely to save the muscles, skin, and bone from rottenness. When I was shown these human petrifactions, I shrank back with disgust and horror. 'Ashes to ashes!' thought I—'Dust to dust!' If this be not dissolution, it is something worse than natural decay—it is treason against humanity, thus to lift up the awful veil which would fain hide its weakness." Later she fortuitously states, "For worlds I would not see a form I loved—embalmed in my heart—thus sacrilegiously handled? Pugh! my stomach turns" (73).

61. The demand for better methods of preservation was seconded by the artists who drew anatomical plates. In the course of Leonardo's studies, which he finally summarized in 750 magnificent plates, he dissected over fifty cadavers, an unusually large number for the time. His venous injection may have served as an inspiration years later to Ruysch and Hunter, according to Robert W. Habenstein and William M. Lamers, *The History of American Funeral Directing* (Milwaukee: Bulfin, 1955), 109. William Harvey, a physician and anatomist, and greatest of physiologists, announced his discovery of the circulation of blood to his students in 1618 (99).

62. Robert Boyle, *Work-diary XXV* ("Loose Experiments, Observations & Notes about the Preservation of Bodyes"), Royal Society, Boyle Papers 27, 219–20.

63. In addition to Boyle, numerous scientists and physicians experimented with putrefaction of solids, including John Pringle, "Some Experiments on Substances Resisting Putrefaction," *Philosophical Transactions* 47 (1749–1750): 480–88, as well as *Observations on the Diseases of the Army* (1752), and George Pearson, *Observations and Experiments for investigating the chymical History of the Tepid Springs of Buxton* [. . .], (1784).

64. Salmon, *Synopsis Medicinæ*, (1681), 392. I must note that the second edition consulted at the Royal College of Surgeons, Edinburgh, was mispaginated: chapter 20, "Of Embalming," com-

menced on page 400 and was followed by 401, 392, 393, 396. An oddly fascinating astrological-medical text, the chapter on embalming was an addition to the greatly expanded second edition.

65. William Wilkins, *A Specimen of Embalming the Dead Intire without Disbowelling* (1690), held at the British Library, microfilm 40469. Wilkins was not a member of the Royal Society. I wish to thank the British Library for its kind attention to the retrieval and copy of this item. Many thanks to Gill Jackson, information officer of the Royal Society.

66. Thomas Greenhill, *NExPoxHΔEIA: or, the Art of Embalming: Wherein is Shewn the Right of Burial, the Funeral Ceremonies, and the several Ways of Preserving Dead Bodies in Most Nations of the World* (London: Printed for the author, 1705). *Nekpokhdeia* or *Nekrokedia* are acceptable English spellings of the title.

67. Greenhill, *NExPoxHΔEIA*, quoted in Habenstein and Lamers, *The History of American Funeral Directing*, 109. The term "undertaker," in its modern sense, goes back to at least 1698, at which date its present usage was recorded in a parish register. The term does not seem to have come into use to designate a funeral contractor until the beginning of the eighteenth century. See Habenstein and Lamers, *The History of American Funeral Directing*, 102–6.

68. For a brief history of the barber-surgeons, see Habenstein and Lamers, *The History of American Funeral Directing*, 100–110.

69. Sidney Young, *Annals of the Barber-Surgeons of London* (London: Blades, 1980), 111–12.

70. In 1745, the surgeons and barbers ended their joint relationship and again became two separate organizations. During the barber-surgeons period they were permitted to be the sole agency for embalming and for performing anatomical dissections in the city of London, although there is no record of any of the bodies for anatomy being embalmed. In 1745, the surgeons formed a separate company of surgeons, which later became the Royal College of Surgeons.

71. My thanks to Robert Folkenflik for the Sheridan reference.

72. According to Olivia Bland in *The Royal Way of Death* (London: Constable, 1986), the undertaker, Mr. Harcourt, produced a bill for £43 9s. 6d. Harcourt laid out the corpse, provided the coffin, the winding sheet, and "two wooden urns covered with lead and lined with silk for ye Bowells," and "six men to move the Body under the canopy" (100).

73. See Sean Silver, *The Mind Is a Collection: Case Studies in Eighteenth-Century Thought* (Philadelphia: University of Pennsylvania Press, 2015), for an evocative interpretation of Reynolds's portrait of William Hunter.

74. For a historical explanation of the Hunter brothers' dissection practices and the acquiring of dead bodies for anatomy instruction, see Anita Guerrini's excellent chapter, "The Value of a Dead Body," in *Vital Matters: Eighteenth Century Views of Conception, Life, and Death* (Toronto: University of Toronto Press, 2012).

75. William Hunter, "The Art of Embalming Dead Bodies," being the substance of a lecture delivered by William Hunter M.D., January 13, 1776, MS0204/1 in the library of the Royal College of Surgeons, London. The original lecture is nonexistent; the Royal College holds the only surviving lecture notes of one of Hunter's students. Many thanks to Tina Craig, librarian, Royal College of Surgeons, for transcribing the lecture notes. Hunter's lectures, excluding the embalming lecture, have been published by Nell Dowd in *Hunter's Lectures of Anatomy* (Amsterdam: Elsevier, 1972). Many of the eighteenth-century embalmers (John Hunter, Richard Mead) also wrote treatises on the bowels and stomach. For a wider reflection on the general manifestations of the period's intellectual, medical, and artistic preoccupations with the viscera and gastric, see Barr et. al., *Bellies, Bowels and Entrails in the Eighteenth Century*.

76. Bob Brier, *The Encyclopedia of Mummies* (New York: Facts on File, 1998), 78.

77. See Brier, *The Encyclopedia of Mummies*, for the details of Hunter's method.

78. In 1782, Hunter constructed a museum between his two homes in London to house his anatomical and natural history collections, now a part of the Hunterian Museum at the University of Glasgow.

79. Alan Friedman, *Fictional Death and the Modernist Enterprise* (Cambridge, UK: Cambridge University Press, 1995), 103. "Despite similar opposition, George IV was embalmed because his funeral was delayed" (*Sunday Times*, 1830).

80. See Chapter 5 in this book for further discussion of Byrne's skeleton. According to Ruth Richardson and Brian Hurwitz in "Jeremy Bentham's Self Image: An Exemplary Bequest for Dissection," *British Medical Journal* 295, no. 6591 (1987): 195–98, John Hunter's portrait contains the skeletal feet of Patrick O'Brien (or Charles Byrne), the celebrated "Irish Giant," "whose skeleton greets one on entering the Hunterian Museum, Royal College of Surgeons of London, and for whose corpse the anatomist John Hunter paid over £500" (197).

81. Baillie's lecture was published in *Transactions of a Society for the Improvement of Medical and Chirurgical Knowledge* 3 (1812): 7–23.

82. James Wardrop, ed., *The Works of Matthew Baillie, M.D. To which is Prefixed an Account of His Life* (London: Longman, Hurst, Rees, et al., 1825), 1:183.

83. Charles Balguy, "An Account of the Dead Bodies of a Man and Woman, which were preserved 49 Years in the Moors in Derbyshire [. . .]," *Philosophical Transactions* 38, (1733): 413–15.

84. In "The Value of a Dead Body," Guerrini notes that catalogs from the 1780s of John Hunter's preparations describe over four hundred specimens, including one hundred dry preparations and the rest "wet," which preserved body parts by immersion in fluid.

85. Deutsch discovers that the illustration was indeed taken from Johnson's lung, and that it was reproduced in twentieth-century medical textbooks. Incidentally, the preserved lung is no longer with Cruikshank's collection and has never been found. The manuscript record of the autopsy is housed at Dr. Johnson's House Museum in London. See Helen Deutsch, *Loving Dr. Johnson* (Chicago: University of Chicago Press, 2005), 29–30.

86. British political interest in Egypt would go on to have a long history. Egypt would find itself under the British from 1882, when it was occupied by British forces during the Anglo-Egyptian War, until 1956, when the last British forces withdrew in accordance with the Anglo-Egyptian agreement of 1954 after the Suez Crisis.

87. John Hadley, "An Account of a Mummy, Inspected at London 1763," *Philosophical Transactions (1683–1775)*, 54 (1764): 1–14.

88. Hadley, "An Account of a Mummy," 1.

89. John Frederick Blumenbach, "Observations on some Egyptian Mummies opened in London," *Philosophical Transactions* 84 (1794): 177–95.

90. Smart Lethieullier's cousin, Captain (later Colonel) William Lethieullier, had a small collection of Egyptian mummies that he eventually donated to the British Museum following Sir Hans Sloane's donation, which formed the British Museum in 1753. In fact, Lethieullier donated his first mummy (no. 6696), which had been in the country thirty-three years, a few years after the founding of the museum. An older mummy (no. 6957) was the property of Nell Gwyn but was not presented to the museum until 1837. William Lethieullier's interest in mummification may have led to Smart's interest in corporeal preservation. For a list of the mummified items, see Warren R. Dawson and P. H. K. Gray, *Catalogue of Egyptian Antiquities in the British Museum* (London, 1968), 1:1–40. See also Alexander Gordon, *An Essay Towards Explaining the Hieroglyphical Figures, on the Coffin of the Ancient Mummy Belonging to Capt. William Lethieullier* (London, 1737), and, for a fascinating description of Egyptian motifs and symbols, Alexander

Gordon, *Essay Towards Explaining the Hieroglyphical Figures, on the Egyptian Mummy, in the Museum of Doctor Mead, Physician in Ordinary to his Majesty* (London, 1737).

91. Public mummy unrollings continued well into the nineteenth century. In 1821, physician Augustus Bozzi Granville FRS unwrapped and dissected an ancient Egyptian mummy, presenting the results at the Royal Society in 1825. He commissioned artist Henry Perry to draw the process in stages, and the drawings were subsequently engraved for publication in *Philosophical Transactions.* Egyptianizing embalming practices continued among the wealthy: Alexander, 10th Duke of Hamilton (d. 1852), for example, was embalmed by physician and mummy expert Thomas "Mummy" Pettigrew and buried in a genuine Egyptian sarcophagus of the Ptolemaic period in a lavish mausoleum (c. 1858) on his family estate in Scotland. Pettigrew still staged public mummy unrolling in the 1830s and 1840s. For more information about the history of the Hamilton Mausoleum, see "Hamilton Mausoleum (South Lanarkshire)," *The Mausolea & Monuments Trust,* available at http://www.mmtrust.org.uk/mausolea/view/501/Hamilton_Mausoleum_South_Lanarkshire (accessed January 28, 2018). Though it is beyond the scope of my project to discuss corpse medicine, I must mention the fact that ground human skull was a component of apothecary shops, consistently prescribed for epilepsy and other ailments in the eighteenth century. Ancient mummy was also ingested. Numerous editions of Samuel Johnson's *Dictionary* listed "mummy" as a medicine. For a fascinating history of the medicinal use of "mummy" in the period (as well as human skull), see Richard Sugg, *Mummies, Cannibals, and Vampires: The History of Corpse Medicine from the Renaissance to the Victorians* (London: Routledge, 2011).

92. Smart Lethieullier to Cromwell Mortimer [Bond Street, London], October 24, 1734, *Royal Society Archives*, EL/L6/45. My special thanks to Nichola Court, library information services, the Royal Society.

93. William Wilkins, *Post Boy*, July 3, 1701.

94. Courtesy of the Historic UK: The History and Heritage Accommodation Guide website, "Hannah Beswick, the Mummy in the Clock," https://www.historic-uk.com/HistoryUK/HistoryofBritain/Hannah-Beswick-The-Mummy-In-The-Clock/.

95. Thomas de Quincey, *Autobiography* (London: Black, 1896), quoted in Dobson, "Some Eighteenth-Century Experiments," 433. Supposedly, Charles White was acting in response to Hannah Beswick's phobia of being buried alive, which had been brought about when her brother, thought to be dead, was found still to be breathing as the coffin was screwed down.

96. When in 1868 it was decided to close the museum, the mummy was buried in an unmarked grave at Harpurhey Cemetery. In the Manchester Museum there still exists the table on which the mummy was displayed.

97. There remains a dispute over the case of Miss Johnson. Sheldon's wife records that her husband barely knew the woman, and on learning that she had donated her body to the hospital, he cast lots with a fellow doctor for her corpse. Others recall that Miss Johnson was indeed Sheldon's mistress and that he preserved her body and kept it in his home as a form of morbid attachment to her. Litten notes that the body was later discovered to be that of Sarah Stone, who worked for Sheldon (but there are disputes over her identity, since Stone survives Sheldon). See Dobson, "Some Eighteenth-Century Experiments," 435–38, and Litten, *The English Way of Death*, 50.

98. See John Kobler, *The Reluctant Surgeon: A Biography of John Hunter* (New York: Doubleday, 1960), 138–39. The 1830 Royal College of Surgeons catalogue described her as "the embalmed body of a female subject aged 24, of the name of Johnson, who died of phthisis in the Lock Hospital, about the year 1775 and left her body for dissection to Mr Sheldon, who was at that time a pupil of that charily. . . . Presented by Mrs Rebecca Sheldon Dec 24th 1808," and "It is stated that 'much camphor was used, that all the arteries and veins were filled with injection, and that spirit of wine was used as well as cam-

phor; that the heart and intestines were taken out, and injected and replaced, as was also the brain.'" *Catalogue of the Contents of the Museum of the Royal College of Surgeons in London*, 9:55.

99. See Dobson, "Some Eighteenth-Century Experiments," 435–38.

100. "The method pursued in its preparation was, principally, that of injecting the vascular system turgid with oil of turpentine and camphorated spirit of wine, and the introduction of powdered nitre and camphor into the cavity of the abdomen, &c." Dobson, "Some Eighteenth-Century Experiments," 55.

101. John Hunter, *Essays and Observations on Natural History, Anatomy, Physiology, Psychology, and Geology* (London: J. Van Voorst, 1861), 1:398–400.

102. Wendy Moore, *The Knife Man: Blood, Body Snatching, and the Birth of Modern Surgery* (Portland, OR: Broadway Books, 2005), 229–30.

103. Horace Walpole, *The Yale Edition of Horace Walpole's Correspondence*, ed. Wilmarth S. Lewis, A. Dale Wallace, and Edwine M. Martz (New Haven, CT: Yale University Press, 1965), 33:535 (letter from Walpole to Lady Ossory, November 4, 1786). At the time of Walpole's letter Grenville was Prime Minister of Great Britain, best known for implementing the Stamp Act and extending the policy to Britain's American colonies, which instigated opposition and contributed to upheavals that led to the American Revolution.

104. Walpole, *Yale Edition*, ed. Wilmarth S. Lewis (1974), 39:9.

105. Walpole, *Yale Edition*, ed. Wilmarth S. Lewis (1961), appendix 8, "Horace Walpole's Will," 30: 377.

106. John Cranch's *Narrative Relating to the Real Embalmed Head of Oliver Cromwell* (London, 1799). See also *The Embalmed Head of Oliver Cromwell in the possession of the Rev. H.R. Wilkinson, exhibited before the Royal Archaeological Institute* (London: The Institute, 1911).

107. The Cromwell Association, http://www.olivercromwell.org. See also H. F. McMains's *The Death of Oliver Cromwell* (Lexington: University Press of Kentucky, 2000) and Lorna Clymer's "Cromwell's Head and Milton's Hair: Corpse Theory in Spectacular Bodies of the Interregnum," *The Eighteenth Century* 40, no. 2 (1999): 91–112.

108. Antonia Fraser, *Cromwell: The Lord Protector* (New York: Knopf, 1973), 681. For an entertaining history of Cromwell's skull in the twentieth century, along with photographs, see Frances Larson's *Severed: A History of Heads Lost and Heads Found* (New York: Liveright, 2014).

109. Though not embalmed, novelist Laurence Sterne's head had journeys of its own: it was buried in St. George's graveyard when Sterne died in 1768; it was disinterred sometime shortly thereafter for anatomists and reburied; and then in 1969 the skull was disinterred and taken, along with a femur, to Coxwold and buried outside the church where he had preached. See Kenneth Monkman and W. G. Day, "The Skull," *The Shandean* 10 (1998): 45–79. Monkman also includes photographs of the disinterment. See Chapter 5 of this book for further discussion.

110. Later, in 1828 Scottish anatomist Robert Knox would be implicated in the Burke and Hare body snatchings, as well as accused of embalming for necrophilic purposes. Knox purportedly displayed the body of Mary Paterson (a young woman Burke and Hare had smothered) "to all and sundry in his school," so that men could come to draw her body. Testimony revealed that "comments had been made upon her physical attributes and that Knox had even had her body preserved in spirits to that he could continue to indulge in necrophiliac voyeurism" (Richardson, 96).

CHAPTER 3

1. Later Charlotte Smith will borrow this description for another fallen woman on her deathbed in *Celestina* (1791). Smith describes Emily Cathcart's skin with similar terms: "Emaci-

ated, and of a delicate fairness, her hands and her face had a transparency that gave an idea of an unembodied spirit, and her dress was such as favoured the deceptions. The blood might almost be seen to circulate in her veins, so plainly did they appear" (Toronto: Broadview, 2004), 523. And disturbingly Mary Shelley's creature in *Frankenstein* (1818) will have transparent skin: "His yellow skin scarcely covered the work of muscles and arteries beneath." While Victor is horrified by the body of his creature, he also deems him "beautiful."

2. Notoriously, the Marquis de Sade is convicted for a sexual crime involving wax, and kicked out of a prison for brandishing wax. He lacerates his victim, then fills her cuts with hot wax. In *The Marquis de Sade: A Life*, biographer Neil Schaeffer summarizes Rose Keller's testimony: "After each whipping, Keller claimed that Sade had dripped hot wax on her wounds, specifying white wax, and also red wax that has a much higher melting point. Sade denied that he had tortured Keller with melted wax." See Schaeffer's *The Marquis de Sade: A Life* (Cambridge, MA: Harvard University Press, 2000), 92. Many biographers note that the "wax" which she claimed had been poured in her abrasions was actually an ointment he applied (doctors did not find scars on her body). In another incident involving wax, while in Sainte-Pélagie prison Sade made "indecent propositions and lascivious gestures with a wax dildo," creating such turmoil that Sade was transferred to two other prisons. For details, see John Franceschina's introduction to *Rape, Incest, Murder! The Marquis de Sade on Stage* (Albany, GA: BeanManor Media, 2015). Like the artificial leg, the dildo was "substituted" for a missing or nonexistent part of the body. Jennifer Van Horn notes that "Made 'of wax horn, leather, and diverse other substances,' as authors of slang dictionaries related, dildos became widely available for purchase in London's 'toy shops and nick nackatories' during the late seventeenth and eighteenth centuries" (370). As with wax anatomical models, most were imported from Italy and France. William Hamilton's personal collection included wax-model phalli (mainly from Italy).

3. See especially Bronfen's "Bodies on Display," in *Over Her Dead Body: Femininity, Death and the Aesthetic* (New York: Routledge, 1992), 95–109.

4. Some helpful sources include Thomas N. Haviland and Lawrence Charles Parish, "A Brief Account of the Use of Wax Models in the Study of Medicine," *Journal of the History of Medicine and Allied Sciences* 25, no. 1 (1970): 52–75; Marina Warner, "Waxworks and Wonderlands," in *Visual Display: Culture Beyond Appearances*, ed. Lynne Cooke and Peter Wollen (Toronto: Bay Press, 1995), 179–201; Anna Maerker, *Model Experts: Wax Anatomies and Enlightenment in Florence and Vienna, 1775–1815* (Manchester, UK: Manchester University Press, 2011); Michelle E. Bloom, *Waxworks: A Cultural Obsession* (Minneapolis: University of Minnesota, 2003); Pamela Pilbeam, *Madame Tussaud: And the History of Waxworks* (London: Hambledon Press, 2003); Mark Sandberg, *Living Pictures, Missing Persons: Mannequins, Museums, and Modernity* (Princeton, NJ: Princeton University Press, 2002); Roberta Panzanelli, *The Color of Life: Polychromy in Sculpture from Antiquity to the Present* (Los Angeles: J. Paul Getty Museum, 2008); Roberta Panzanelli and Julius von Schlosser, *Ephemeral Bodies: Wax Sculpture and the Human Figure. With a Translation of Julius Von Schlosser's "History of Portraiture in Wax"* (Los Angeles: J. Paul Getty Museum, 2008); Samuel Alberti, "Wax Bodies: Art and Anatomy in Victorian Medical Museums," *Museum History Journal* 2, no. 1 (2009): 7–36; Anna Maerker, *Model Experts: Wax Anatomies and Enlightenment in Florence and Vienna, 1775–1815* (Manchester, UK: Manchester University Press, 2011); Elizabeth Stephens, *Anatomy as Spectacle: Public Exhibitions of the Body from 1700 to the Present* (Liverpool: Liverpool University Press, 2011); Ludmilla Jordanova, *Sexual Visions: Images of Gender In Science and Medicine Between the Eighteenth and Twentieth Centuries* (Madison: University of Wisconsin Press, 1993); Joanna Ebenstein, *The Anatomical Venus: Wax, God, Death & the Ecstatic* (London: Thames & Hudson, 2016); Lucia Dacome, *Malleable*

Anatomies: Models, Makers, and Material Culture in Eighteenth-Century Italy (Oxford: Oxford University Press, 2017); Davide Stefanacci, *The Vision of Death: Wax Sculpture in the Seventeenth Century* (Scotts Valley, CA: CreateSpace, 2018).

5. By the mid-eighteenth century, many anatomical museums (such as Surgeons' Hall) charged admission for the public to view body parts, both floating wet specimens in jars and wax models of the body's interior.

6. Susan M. Stabile, *Memory's Daughters: The Material Culture of Remembrance in Eighteenth-Century America* (Ithaca, NY: Cornell University Press, 2004), 95.

7. Some Texts from Early Modern Philosophy website, "René Descartes," accessed July 22, 2021, https://www.earlymoderntexts.com/authors/descartes.

8. "René Descartes."

9. Neil Hertz, "Dr. Johnson's Forgetfulness, Descartes' Piece of Wax," *Eighteenth-Century Life* 16 (1992): 175.

10. Elizabeth Hallam and Jenny Hockey, *Death, Memory and Material Culture* (Oxford: Berg, 2001), 29.

11. Stabile, *Memory's Daughters*, 95.

12. John Locke, *An Essay Concerning Human Understanding, and A Treatise on the Conduct of the Understanding* (1760), Book II, Chapter 29, Section 3.37.

13. Laurence Sterne, *Tristram Shandy* (London: Penguin, 2003), 71. Editor Melvyn New notes that "brass-jack" could refer to a counterfeit coin. The influence of Locke's *Essay* on *Tristram Shandy* has seen much attention. McMaster argues for Burton's *The Anatomy of Melancholy* as a more congenial model.

14. "The rise of the Wax Chandlers Company corresponded with the proliferation of trade and parish fraternities in London, and the increasingly elaborate religious ceremonials of the late C14 and C15. 'Proper and seemly funerals' called for enormous tapers and churches and altars were lit all day every day and ornamented with numerous wax images." Some sources date the origins of the fraternity to the twelfth century. For a detailed history of the livery, see http://www.waxchandlers.org.uk/index.php.

15. Luke Syson, "Polychrome and Its Discontents: A History," in *Like Life: Sculpture, Color, and the Body*, ed. Kamilah Foreman, Elizabeth Franzen, et al. (New York: Metropolitan Museum of Art, 2018), 26.

16. Elizabeth Kowaleski Wallace, "Recycling the Sacred: The Wax Votive Object and the Eighteenth-Century Wax Baby Doll," in *The Afterlife of Used Things: Recycling in the Long Eighteenth Century* (New York: Routledge, 2014).

17. John Evelyn's diary, April 4, 1672 (*Diaries*, 3:612), quoted in Syson, 26.

18. Syson, 26. The use of wax effigies is an ancient practice. "Since noble Romans carried the images of the dead at funerals, they were necessarily made of light materials such as wax" (82). See Matthew Craske, *The Silent Rhetoric of the Body: A History of Monumental Sculpture and Commemorative Art in England, 1720–1770* (New Haven, CT: Yale University Press, 2008).

19. Deborah Lutz, *Relics of Death in Victorian Literature and Culture* (Cambridge, UK: Cambridge University Press, 2015), 195n61.

20. Craske, *The Silent Rhetoric of the Body*, 240–41. Craske describes further that "the figures had been displayed in the main hall of Buckingham House beneath an allegory of the Duke and Duchess painted by Antonio Belluchi in 1722" (241).

21. Syson, "Polychrome and Its Discontents," 26.

22. In Appendix V in the Oxford University Press edition (1988) of Burney's *Cecilia, or Memoirs of an Heiress* (1782), Peter Sabor notes that "It was a sign of wealth to be able to turn

night into day. Candles were expensive; fine wax candles such as Cecilia's hosts would use cost about 3 shillings a pound, and a party given at home might cost several guineas in candles alone" (955).

23. *The Diary of Samuel Pepys*, "Thursday 15 December 1664," accessed March 5, 2019, https://www.pepysdiary.com/diary/1664/12/15/.

24. Burney's novel *Cecilia* also mentions "Turner's cerate," a stiff ointment made of calamine and wax composed by Daniel Turner and popular into the nineteenth century.

25. *The lady's delight, or accomplish'd female instructor: being a very useful companion for ladies, gentlewomen, and others* [...] (London: Printed for James Hodges), 1740.

26. Charlotte Smith's *Ethelinde, or The Recluse of the Lake* (Richmond, VA: Valancourt Books, 2017), 445.

27. Julie Park, *The Self and It: Novel Objects in Eighteenth-Century England* (Stanford, CA: Stanford University Press, 2010), xv.

28. The 1773 catalog for Cox's Museum notes the expense of admission as 5s. 3d. (410n76). At the end of the catalog, the anonymous editor (perhaps Cox) writes, "Mr. Cox having been informed that some of the Nobility and Gentry, who have not as yet done him the honour of inspecting his Museum, think he has set too high a price on his tickets of admission, he takes the liberty of soliciting their attention to the very different circumstances that distinguish this from every other Exhibition."

29. See Anonymous, *A Descriptive Catalogue of the Sevveral Superb and Magnificent Pieces of Mechanism and Jewellery, Exhibited in the Museum, at Spring Gardens, Charing Cross* (London, 1773), 28–29. For a helpful discussion of Burney's novels and automated pleasures, see Julie Park, "Pains and Pleasures of the Automaton: Frances Burney's Mechanics of Coming Out," *Eighteenth-Century Studies* 40, no. 1 (2006): 36. The catalog is assumed to have been written by Cox.

30. Frances Burney, *Evelina, or the History of a Young Lady's Entrance into the World* (Oxford: Oxford University Press, 1982), 76.

31. Anonymous, *Extraordinary, and unparallel'd curiosities, composed of wax, (so near the life, that they cannot be excelled by art) in seven several scenes* (London, 1664). The waxworks are noted to be at the "Booth next to Cloth-Fair-Gate in *Smithfield*."

32. Laura Engel, "Epilogue: Body Double—Katharine Hepburn at Madame Tussauds," in *Intimacy and Celebrity in Eighteenth-Century Literature and Culture: Public Interiors* (London: Palgrave Macmillan, 2018), 295. The handbills for the house of wax are held at the Bodleian Libraries. "While some accounts say the waxworks—which, according to the City of London website remained at the site until 1816—were taken over by a Chancery Lane surgeon named Clarke after Mrs Salmon's death (and by his wife after his death), it is also suggested that at some point they moved to Water Lane in east London where they were ruined by thieves. Whatever its fate, it's generally accepted that the famous waxworks were visited by the likes of James Boswell and artist William Hogarth." See Exploring London, "Lost London—Mrs Salmon's Waxworks," https://exploring-london.com/2013/04/19/lost-london-mrs-salmons-waxworks, as well as Richard Altick's *The Shows of London* (Cambridge, MA: Belknap Press, 1978).

33. Park, *The Self and It*, 95.

34. In Charlotte Smith's *Desmond* (1792), the narrator mentions "a hired or wax hermit in some of our gardens" (Peterborough, UK: Broadview, 2001), 118.

35. Anna Meades, *The History of Sir William Harrington. Written Some Years Since, And revised and corrected by the late Mr. Richardson* [...], 4 vols., 3:4 (London: 1772). Meades corresponded with Richardson, but he passed on publishing her novel. Though the novel is attributed to Richardson on the title page, Meades is acknowledged by scholars as the author.

36. The object notes state, "This seems an unusual addition, but its inclusion is typical of such displays where 'exotic' objects were often displayed alongside the waxes to create an atmosphere." The print for *Part of Aubin's Cabinet* was etched by Stefano Superchi and is held at the Victoria and Albert Museum, E.2721-2016.

37. Kate Berridge, *Madame Tussaud: A Life in Wax* (New York: Harper Perennial, 2007), 4–5.

38. Marjan Sterckx, "Pride and Prejudice: Eighteenth Century Women Sculptors and Their Material Practices," in *Women and the Material Culture of Death*, ed. Maureen Daly Goggin and Beth Fowkes Tobin (Farnham, UK: Ashgate, 2013), 293.

39. Marina Warner, *Phantasmagoria: Spirit Visions, Metaphors, and Media into the Twenty-First Century* (Oxford: Oxford University Press, 2008), 40.

40. For a helpful early work on wax modelers, see E. J. Pyke, *A Biographical Dictionary of Wax Modellers* (New York: Clarendon Press, 1973).

41. Sleeping Beauty was cast by Curtius in 1765. At the turn of the century, her clockwork breast was electrified. She was recast in 1925 after being destroyed in a fire. In 1802 Curtius's Curiosity Cabinet toured Britain for the first time. Marie continued the work following the death of Curtius in 1794. For a detailed history of waxworks, see Warner's *Phantasmagoria.*

42. Joanna Ebenstein, *The Anatomical Venus: Wax, God, Death & the Ecstatic* (London: Thames & Hudson, 2016), 188–89.

43. Warner, *Phantasmagoria*, 51.

44. Warner, *Phantasmagoria*, 49.

45. See Luuc Kooijmans, *Death Defied: The Anatomy Lessons of Frederik Ruysch*, trans. Diane Webb (Boston: Brill, 2010).

46. Steven Jay Gould, in his book *Finders, Keepers: Eight Collectors*, 15–18. Ruysch's tableaux did not stand the test of time and exist only in book illustrations. His wet preparations were gifted to several museums, including St. Petersburg's Kunstkammer.

47. *Journal Book of the Royal Society*, 14:153–54, quoted in John H. Appleby, "Human Curiosities and the Royal Society, 1699–1751," *Notes and Records of the Royal Society of London* 50, no. 1 (1996): 14. For visual examples of wax preservation, see http://morbidanatomy.blogspot.com/.

48. John Abernethy, "Observations on the foramina Thebesii of the heart," *Philosophical Transactions* 88 (1798): 105. The Wellcome Institute holds several drawings taken from these wet, waxed specimens, https://wellcomelibrary.org/item/b20457042.

49. See John H. Teacher, ed., *Catalogue of the Anatomical and Pathological Preparations of Dr. William Hunter* (Glasgow: James MacLehose and Sons, 1900), xxi–lxxxi.

50. See Martin Kemp and Marina Wallace, *Spectacular Bodies* (London: Hayward Gallery, 2000), 11–19.

51. The few female bodies that appeared in anatomical atlases before the mid-eighteenth century borrowed from the conventions of the female nude in Western art. See Marcia Nichols, "Venus Dissected: The Visual Blazon of Mid-18th Century Medical Atlases," in *Sex and Death in Eighteenth-Century Literature*, ed. Jolene Zigarovich (New York: Routledge, 2013), 103–26. Rebecca Marie Messbarger recognizes the unique case of role reversal in the male-dominated field of anatomy: that of Anna Morandi. Scholars generally agree that Morandi was the only female wax modeler of male genitalia. See Messbarger, *The Lady Anatomist: The Life and Work of Anna Morandi Manzolini* (Chicago: University of Chicago Press, 2010), and Lucia Dacome, *Malleable Anatomies: Models, Makers, and Material Culture in Eighteenth-Century Italy* (Oxford: Oxford University Press, 2017).

52. See Caroline Grigson, "'An universal language': William Hunter and the Production of *The Anatomy of the Human Gravid Uterus*, 59–80, Anne Dulau Beveridge, "The Anatomist and the Artists: Hunter's Involvement," 81–96, and Stuart W. McDonald, "William Hunter's anatomical and pathological specimens," 97–114, all in E. Geoffrey Hancock et al., eds., *William Hunter's World: The Art and Science of Eighteenth-Century Collecting* (New York: Routledge, 2017). John Hunter also commissioned numerous wax models of the human anatomy (including male genitalia). Modeling in wax, plaster, and clay was necessary for the anatomist to capture the actual workings and demystify the secrets of the human body. Anita Guerrini notes that John Hunter also "appears especially enamoured of the technique of the écorché, where a corpse is flayed to reveal the structure of the muscles" ("Anatomists" 256). Fictional references to the écorché are rare, but in Jonathan Swift's satire *A Tale of the Tub* (1704), the narrator contrasts philosophical wisdom that seeks to plumb the depths with a wisdom of the surface. Using an anatomical example, he writes, "I do here think it fit to inform the reader that in such conclusions as these, reason is certainly in the right, and that in most corporeal beings which have fallen under my cognizance, the *outside* hath been infinitely preferably to the *in*: whereof I have been further convinced from some late experiments. Last week I saw a woman *flayed*, and you will hardly believe how much it altered her person for the worse" (*A Tale of a Tub and Other Works*, [Oxford: Oxford University Press, 1999], 84). Notably, the flesh endlessly flayed and punished by the Church is now flayed in the name of science.

53. Courtney Wennerstrom remarks, "The waxen figure—when viewed in relationship to international histories of dissection, medical moulage, obstetrics and anatomia plastics—allows for feminist and queer revisions of sexual categories" ("Cosmopolitan Bodies and Dissected Sexualities: Anatomical Mis-Stories in Ann Radcliffe's *Mysteries of Udolpho," European Romantic Review* 16, no. 2 (2005): 196.). Bronfen's feminist reading links the rise of the anatomical model with the popular folktale Snow White, which depicts the dwarves placing Snow White's comatose body in a glass coffin, offering itself to an aesthetic gaze. Bronfen aligns this image with Celemente Susini's and Giuseppe Ferrinni's wax Venuses, nude, dissected, and preserved in glass cases at La Specola. Bronfen, as well as more recent scholars such as Ebenstein, convincingly reads the Venuses as pure "erotic desire" and auto-icons. Bronfen notes that the "centre and origin of human life as signified by the interior of the feminine body" are represented by the wax cast of the cadaver (99). For her insightful discussion of Snow White and the Venuses, see 99–107.

54. Ebenstein, *Anatomical Venus*, 15. La Specola opened to the public in Florence on February 21, 1775. Ebenstein notes, "The first full-sized female instructional anatomical wax models began to be created in the early eighteenth century. French surgeon and anatomist Guillaime Desnous publicly exhibited a dissectible wax woman featuring a newborn child with the umbilical cord still attached" (37).

55. See A. W. Bates, "'Indecent and Demoralising Representations': Public Anatomy Museums in mid-Victorian England," *Medical History* 52, no. 1 (2008): 1–22. Estimate cited by Bates in Oliver Elton, ed., "History of the anatomical museum," in *A miscellany presented to John Macdonald Mackay, LL.D. July, 1914* (Liverpool: Liverpool University Press, 1914), 305. Bates charts the decline of public anatomy museums in the nineteenth century and the effect of the Obscene Publications Act on their closures.

56. Samuel Alberti, *Morbid Curiosities: Medical Museums in Nineteenth-Century Britain* (Oxford: Oxford University Press, 2011), 3.

57. See Andrew Cunningham, *The Anatomist Anatomis'd: An Experimental Discipline in Enlightenment Europe* (London: Routledge, 2016).

58. See Ellen Adams, "Defining and Displaying the Human Body: Collectors and Classics During the British Enlightenment," *Hermathena* 187 (Winter 2009): 65–97.

59. *A descriptive catalogue (giving a full explanation) of Rackstrow's Museum: consisting of a large and very valuable collection of most curious anatomical figures, and real preparations; also figures resembling life . . . ; to be seen at No. 197, Fleet-Street* [. . .] (London, 1782). For an early critical assessment of Rackstrow's reliance on spectacle, see Altick, *The Shows of London*, 55–56. Matthew Craske's "'Unwholesome' and 'Pornographic': A Reassessment of the Place of Rackstrow's Museum in the Story of Eighteenth-Century Anatomical Collection and Exhibition," *Journal of the History of Collections* 23, no. 1 (2011): 75–99, is critical of historians who find Rackstrow's collections quack science and pure spectacle. See also The Wellcome Library for helpful information about the collection and a copy of the catalog: https://wellcomecollection.org/works/cu8ewdtk.

60. Richard Barnett, "Lost Wax: Medicine and Spectacle in Enlightenment London," *Lancet* 372, no. 9636 (2008): 366.

61. Craske, "'Unwholesome' and 'Pornographic,'" 92. Shapira notes that Benjamin Rackstrow's longtime partner and eventual successor, midwife Catherine Clarke, ran a lying-in facility adjacent to the museum and offered her teaching services to "Any gentlewoman desirous of a thorough knowledge of midwifery, both theory and practice" (149).

62. The Revolutionary and Napoleonic Wars may have shifted the culture's interest to the theater of war, but as Barnett notes, "This public taste for the bizarre continued well into the 19th century. In 1825, the 'Florentine Venus,' a wax model of a girl with detailed internal organs, was shown in pubs in Soho, and in 1839 the first anatomically correct figure of a man, complete with removable buttocks and tactfully veiled genitals, went on display in a house on Margaret Street" (367). Graphic anatomical models were even shown at the 1851 Great Exhibition. For a helpful discussion of Victorian anatomical waxworks, see Kristin D. Hussey, "Seen and Unseen: The Representation of Visible and Hidden Disease in the Waxworks of Joseph Towne at the Gordon Museum," *Interdisciplinary Studies in the Long Nineteenth Century* 24 (2017), https://doi.org/10.16995/ntn.787.

63. Wendy Moore, *The Knife Man: Blood, Body Snatching, and the Birth of Modern Surgery* (New York: Broadway Books, 2005), 230.

64. Horace Walpole, *The Yale Edition of Horace Walpole's Correspondence*, ed. Wilmarth S. Lewis, A. Dale Wallace, and Edwine M. Martz (New Haven, CT: Yale University Press, 1965), 33:535 (letter from Walpole to Lady Ossory, November 4, 1786).

65. Life masks were popular as well. In 1669, Samuel Pepys "took the fashionable step of having a life mask made." See the Royal Society site for its description at https://blogs.royalsociety.org/history-of-science/2013/08/06/death-masks/.

66. An image of Newton's mask can be found at the Royal Society website: https://blogs.royalsociety.org/history-of-science/2013/08/06/death-masks/.

67. Cruickshank supervised the taking of the mask, and the sculptor James Hoskins transformed it into a bust. The death mask is now housed at the National Portrait Gallery, London. See Deutsch, *Loving Dr. Johnson*, 51–53.

68. Haydon took Keats's life mask, noting that Keats endured "hours of discomfort." See Nicholas Roe's *John Keats: A New Life* (New Haven, CT: Yale University Press, 2012), 130. Joseph Severn painted Keats on his deathbed, and Gheradi made Keats's death mask and body casts. Deborah Lutz notes that "One of the masks made from this cast can be found at the Keats-Shelley Museum, in Rome, another at the Keats House Museum, London, and a third at the National Portrait Gallery, London," 191n12. See also Andrew Motion, *Keats* (New York: Farrar, Straus and Giroux, 1998); Thomas McFarland, *The Masks of Keats: The Endeavour of a Poet* (Oxford: Oxford University Press, 2000); and Stanley Plumly, *Posthumous Keats: A Personal Biography* (New York: W. W. Norton, 2008).

69. A helpful early work on wax portraits is David Robin Reilly, *Portrait Waxes: An Introduction for Collectors* (London: Batsford, 1953).

70. For a thorough listing of medallions from the period, see Pyke's *Dictionary of Wax Modellers*. Wax portraits could be quite meticulous, such as that of Mrs. Elizabeth Best (c. 1780), done in polychrome by Samuel Percy, one of the most famous artists to work in the medium. In the Best portrait, additional sheets of sculpted wax and chasing were used for incredible detail and dimension.

71. *Lord Nelson* (c. 1806, Catherine Andras) A.41–1970, and *Princess Charlotte* (c. 1817, artist unknown). Both are held by the Victoria and Albert Museum, London.

72. See the entry for *Lord Nelson* at http://collections.vam.ac.uk/item/O77802/lord-nelson-relief-andras-catherine/. Andras also modeled the wax head and left hand of Nelson for the effigy housed in Westminster Abbey. https://www.westminster-abbey.org/abbey-commemorations/commemorations/horatio-viscount-nelson.

73. James Daybell, *The Material Letter in Early Modern England: Manuscript Letters and the Culture and Practices of Letter-Writing, 1512–1635* (New York: Palgrave Macmillan, 2012), 106.

74. In *Mourning Dress: A Costume and Social History* (1983), Lou Taylor mentions that the portrait passed into the family of the Earl of Shelbourne, through Mrs. Salusbury's daughter, Mrs. Thrale, later Mrs. Piozzi.

75. See Philippe Ariès, *The Hour of Our Death*, trans. Helen Weaver (New York: Knopf, 1981). Numerous examples are now housed and displayed in the undercroft of Westminster Abbey. See "Wax Works at Westminster Abbey," *Scientific American* 80, no. 10 (1899): 148; Anthony Harvey and Richard Mortimer, eds., *The Funeral Effigies of Westminster Abbey* (Woodbridge, UK: Boydell Press, 2003). Deborah Lutz notes that "many recumbent figures at Westminster Abbey have countenances based on death masks, such as that of King Henry VII, Elizabeth I, and Elizabeth of York," *Relics of Death in Victorian Literature and Culture*, 195n62.

76. David Jones, *Gothic Effigy: A Guide to Dark Visibilities* (Manchester, UK: Manchester University Press, 2018), 73.

77. Llewellyn, *Art of Death*, 55. In *The Hour of Our Death*, Ariès mentions that "The last effigies to be used at a funeral were those of the duke and duchess of Buckingham, who died in 1735 and 1743, respectively."

78. Marquis de Sade, *Letters*, quoted in Schaeffer, 199–200. Sade names the artist Gaetano Giulio Zumbo as "Zummo" (which was common). Zumbo's anatomical models are now housed in the Museum of Zoology and Natural History (La Specola), Florence.

79. For a helpful discussion of Zumbo, *Udolpho*, and aesthetics, see Ann Louise Kibbie, "Realism and Decay in Wax," *Configurations* 25, no. 2 (2017): 165–87.

80. Laqueur includes a fascinating narrative of Voltaire's deathbed and corpse: "His death and the question of what would become of his body was therefore of national and international interest, the final battle of a long, portentous war: How would the great enemy of religious superstition die?" He also includes the full image of Curtius's wax relief before it was damaged. See Thomas W. Laqueur's *The Work of the Dead: A Cultural History of Mortal Remains* (Princeton, NJ: Princeton University Press, 2015), 189–203.

81. David Jones points to Madame Tussaud's uncle Dr. Philippe Curtius as an influence on Ann Radcliffe's own Gothic productions. Having opened his "Caverne des Grands Voleurs" in 1782 in Paris, Curtius exhibited "the figures of famous criminals and also, later, effigies of the royal family who had been executed." Bringing together of thieves and murderers influences Schiller's *Die Räuber* (1781), which draws Radcliffe to write Emily's encounter with the wax figurine in *Udolpho*. See Jones, *Gothic Effigy*, 75.

82. Ann Radcliffe's *A Sicilian Romance* (Oxford: Oxford University Press, 1998), 177. Later, Matthew Lewis borrows this wax plot for *Venoni or, The novice of St. Mark's: a drama, in three acts* (1808). Here, Venoni has been tricked into thinking his love Josepha has died in a convent, but in fact he witnessed the burial of a wax effigy in her place.

83. Diane Long Hoeveler, *Gothic Feminism: The Professionalization of Gender from Charlotte Smith to the Brontës* (University Park, PA: Pennsylvania State University Press, 1998), 68.

84. Wennerstrom, "Cosmopolitan Bodies and Dissected Sexualities," 196.

85. Ann Radcliffe, *The Mysteries of Udolpho* (Oxford: Oxford University Press, 1998), 248–49.

86. Terry Castle, "Introduction," in Radcliffe, *The Mysteries of Udolpho*, xxii.

87. Kibbie, "Realism and Decay in Wax," 166.

88. As I discuss in Chapter 5, in Radcliffe's *The Romance of the Forest*, La Motte *does* look twice upon a Gothic horror: a human skeleton. The narrator describes that "La Motte went towards [the door], and as he passed, looked into the recess. Upon the ground within it, stood a large chest, which he went forward to examine, and lifting the lid, he saw the remains of a human skeleton. Horror struck upon his heart, and he involuntarily stepped back. During a pause of some moments, his first emotions subsided. That thrilling curiosity, which objects of terror often excite in the human mind, impelled him to take a second view of this dismal spectacle" (53–54).

CHAPTER 4

1. Elaine Scarry argues that for Donne, "Touch is the model of all the senses," and that he "insists on the obligation to touch the human body, whether acutely alive or newly dead." See "Donne: 'But yet the body is his booke,'" in *Literature and the Body: Essays on Populations and Persons*" (Baltimore, MD: Johns Hopkins University Press, 1988), 88.

2. Samuel Richardson, *Clarissa, or The History of a Young Lady* [1751], 3rd ed., 8 vols. (London: printed for S. Richardson and sold by John Osborn, 1747–1748). All subsequent quotes are from this edition.

3. Kathleen M. Oliver, "'With My Hair in Crystal': Mourning Clarissa," *Eighteenth-Century Fiction* 23, no. 1 (2010): 35. See also *Narrative Mourning: Death and Its Relics in the Eighteenth-Century British Novel* (Lewisburg, PA: Bucknell University Press, 2020) for Oliver's fuller examination of relics and the novel.

4. In *Consumption and Literature: The Making of the Romantic Disease* (New York: Palgrave Macmillan, 2006), Clark Lawlor connects Clarissa's meticulous preparations for death with the seventeenth-century tradition of the good death, and the various inscriptions and devices on her self-designed coffin lid with "John Donne's notoriously pious and eccentric preparation for his death from consumption," 65.

5. Ralph Houlbrooke, in *Death, Religion, and the Family in England, 1480–1750*, and several others have noted Richardson's use of both Puritan and sacramental traditions in the descriptions of Clarissa's death. I do not disagree with this; in places her rhetoric invokes these traditions. Foremost, though, I interpret her meticulous death preparations as a renegotiation of the role of religiosity in dying and death in the period, and her control over her own corpse and its aesthetic presentation as individual, commemorative, and modern.

6. Deborah Lutz, *Relics of Death in Victorian Literature and Culture* (Cambridge, UK: Cambridge University Press, 2015), 21.

7. Peter Marshall, *Beliefs and the Dead in Reformation England* (Oxford: Oxford University Press, 2002), 265.

8. Alexandra Walsham notes that "The dual status of Anglo-Saxon and Norman monarchs as sovereigns and saints made it difficult, indeed politically dangerous, to slight and denigrate their remains." See "Skeletons in the Cupboard: Relics after the English Reformation" for an excellent study of the "survival and apparent resurgence" of the cult of relics, in *Past and Present* 206, no. S5 (2010): 139.

9. Walsham claims that "there are remarkably few cases in which Protestants are said to have imputed thaumaturgic powers to the human remains they gathered," and that "miraculous cures effected by Protestant relics are conspicuous in their absence," "Skeletons in the Cupboard," 137.

10. Walsham, "Skeletons in the Cupboard," 22. Walsham sees this as a religious phenomenon as assumptions persisted about the "physical body as a receptacle for the soul," 143.

11. Edward Gibbon, *The History of the Decline and Fall of the Roman Empire* (London, 1776–1788), 3:xxviii.

12. Marcia Pointon, *Brilliant Effects: A Cultural History of Gem Stones and Jewellery* (New Haven, CT: Yale University Press, 2010), 295.

13. Walsham lists notable pious relics and their display in "profane" exhibitions. See "Skeletons in the Cupboard," 141–42.

14. *Oxford English Dictionary Online* (Oxford University Press, updated June 2021), s.v. "relic," accessed April 29, 2019, https://www-oed-com.ccl.idm.oclc.org/view/Entry/161910?rskey=WFoyPH&result=1.

15. Phillippy, *Women, Death and Literature*, 8.

16. Robin Jaffee Frank, *American Portrait and Mourning Miniatures* (New Haven, CT: Yale University Press, 2000). Speaking to the popularity of American posthumous miniatures, Frank remarks, "At the end of the eighteenth century and the beginning of the nineteenth, 'morbid' depictions of the 'desirable corpse' were first detectable," 147. See also Graham Reynolds, *English Portrait Miniatures* (Cambridge, UK: Cambridge University Press, 1988); Julie Aronson and Marjorie E. Wieseman, *Perfect Likeness: European and American Portrait Miniatures from the Cincinnati Art Museum* (New Haven, CT: Yale University Press, 2006); Stephen Lloyd, *Portrait Miniatures from the National Galleries of Scotland* (Edinburgh: National Galleries of Scotland, 2006); Cory Korkow, *British Portrait Miniatures: The Cleveland Museum of Art* (London: Giles, 2013); Hanneke Grootenboer, *Treasuring the Gaze: Intimate Vision in Late Eighteenth-Century Eye Miniatures* (Chicago: University of Chicago Press, 2013).

17. Constance Classen, *The Deepest Sense: A Cultural History of Touch* (Urbana: University of Illinois Press, 2012), 40.

18. Classen reminds us that when William the Conqueror fought and defeated Harold of Wessex in 1066, he was wearing a bag of relics around his neck. See *The Deepest Sense*, 39.

19. Patricia Fumerton argues that this relic making provides a pattern for personal concerns. See *Cultural Aesthetics: Renaissance Literature and the Practice of Social Ornament* (Chicago: University of Chicago Press, 1993), 8–11.

20. The locket is engraved with the inscription "Hair of Mary Tudor Queen of France cut from her head September 6 1784 when her tomb at St. Edmundsbury was opened H.W." See the Lewis Walpole Library site for images: https://libsvcs-1.its.yale.edu/strawberryhill/oneitem.asp?id=662.

21. Ginny Redington Dawes and Olivia Collings, *Georgian Jewellery 1714–1830* (Woodbridge, UK: ACC Art Books, 2007), 158.

22. For example, in his will, William Shakespeare (d. 1616) made a bequest that certain friends receive rings valued at twenty-six shillings. See Margaret Hunter, "Mourning Jewellery: A Collector's Account," *Costume* 27, no. 1 (1993): 9.

23. Frances Parthenope Verney and Margaret Maria Williams-Hay Verney, *Memoirs of the Verney Family* (London: Longmans, Green, 1892), 4:327.

24. Jonathan Swift quickly had the clothes and ring sold for money, bringing a total value of less than five pounds. See Martha F. Bowden, "Introduction," in *The Reform'd Coquet, Familiar Letters Betwixt a Gentleman and a Lady, and The Accomplish'd Rake*, ed. Martha F. Bowden (Lexington: University Press of Kentucky, 1999), ix–xlvi.

25. Pepys also provided a list of who would receive mourning cloth when he died. Used to make mourning clothing, armbands, and hatbands, the cloth was given to relatives, servants, and dependents. For more on the significance of mourning rings and other jewelry, see Shirley Bury, *An Introduction to Sentimental Jewellery* (London: HMSO, 1985); Diana Scarisbrick, *Rings: Symbols of Wealth, Power and Affection* (New York: Abrams, 1993); Maureen DeLorme, *Mourning Art & Jewelry* (Atglen, PA: Schiffer Publishing, 2004); Christiane Holm, "Sentimental Cuts: Eighteenth-Century Mourning Jewelry with Hair," *Eighteenth-Century Studies* 38, no. 1 (2004): 139–43; Susan M. Stabile, *Memory's Daughters: The Material Culture of Remembrance in Eighteenth-Century America* (Ithaca, NY: Cornell University Press, 2004); Mary Brett, *Fashionable Mourning Jewelry, Clothing, & Customs* (Atglen, PA: Schiffer Publishing, 2006); Dawes and Collings, *Georgian Jewellery: 1714–1830*; Helen Sheumaker, *Love Entwined: The Curious History of Hairwork in America* (Philadelphia: University of Pennsylvania Press, 2007); Anne Louise Luthi, *Sentimental Jewellery* (Buckinghamshire, UK: Shire Publications, 2007); Marcia Pointon, *Brilliant Effects: A Cultural History of Gem Stones and Jewellery* (New Haven, CT: Yale University Press, 2010); Sarah Nehama and Anne E. Bentlry, *In Death Lamented: The Tradition of Anglo-American Mourning Jewelry* (Boston: Massachusetts Historical Society, 2012); Diana Scarisbrick, *Rings: Jewelry of Power, Love and Loyalty* (London: Thames & Hudson, 2014); Rachel Church, *Rings* (London: Thames & Hudson, 2017).

26. The Duke of Richmond and Lord Frederick Campbell were Walpole's executors. See Appendix 8, Horace Walpole's Will, *Horace Walpole's Correspondence: Yale Edition*, the Lewis Walpole Library at Yale, 375. http://images.library.yale.edu/hwcorrespondence/.

27. James Hervey, *Meditations Among the Tombs: In a Letter to a Lady*, 12th ed. (Dublin, 1758), 2:24.

28. David Nokes, *Samuel Johnson: A Life* (New York: Henry Holt, 2009), 134.

29. Daniel Defoe, *Moll Flanders* (London: Penguin, 1989), 272.

30. Daniel Defoe, *The Life and Actions of Moll Flanders* (London: T. Read, 1723), 189.

31. Tobias Smollett, *The Adventures of Peregrine Pickle* (Oxford: Oxford University Press, 1969), 32.

32. Ariane Fennetaux, "Fashioning Death/Gendering Sentiment: Women and the Material Culture of Mourning in Late Eighteenth-Century Europe," in *Women and the Material Culture of Death*, ed. Maureen Daly Goggin and Beth Fowkes Tobin (New York: Routledge, 2016), 31.

33. For examples of eighteenth-century mourning jewelry orders and a discussion of these archives, see Hunter, "Mourning Jewellery," and Fennetaux, "Fashioning Death/Gendering Sentiment."

34. Fennetaux, "Fashioning Death/Gendering Sentiment," 32.

35. Lou Taylor, *Mourning Dress: A Costume and Social History* (New York: Routledge, 1983), 188.

36. By the nineteenth century, mourning jewelry iconography distinctly shifted from the sentiment embodied by the female mourner. Jewelry carved in jet was distinctly worn by women,

and hair jewelry was often homemade by middle-class women. The Victorians mainly rejected the man of sentiment, and the practice of bodily adornments worn by men diminished. In *Mourning Dress*, Lou Taylor remarks, "By the 1870s and 1880s mourning jewellery could be really huge, with great chains and enormous pendants. During the 1890s art nouveau swirls and foliage can be seen," 187.

37. Victorian novels set earlier in the century frequently mention the bequeathing of money to purchase mourning jewelry, or the actual wearing of such (see Lutz, *Relics of Death*, for this discussion and an excellent reading of Emily Brontë's *Wuthering Heights*). In Charlotte Brontë's *Jane Eyre*, thirty guineas are bequeathed to Mr. Rivers and his two sisters to buy mourning rings. In Anne Brontë's *The Tenant of Wildfell Hall*, Helen Graham wears a hair chain on which a watch is suspended. In Anthony Trollope's *The Last Chronicle of Barset*, the narrator states that following the death of his wife, Adolphus Crombie is consoled by a mourning ring with his wife's hair. See Hunter, "Mourning Jewellery," 11.

38. Donne's speaker anticipates that the grave digger will discover the lover's hair about the bone of the speaker's forearm, thinking the intermingled remains are of a couple using the hair as a cipher so that at Judgment Day their souls might meet at the grave. "The Relique" famously invokes the Catholic veneration of saints relics in a post-Reformation context; the formerly Catholic Donne indeed could have been charged with heresy.

39. The Victorians sometimes used bones and teeth for mementos to demonstrate their grief. See Taylor, *Mourning Dress*, 200. I agree with Taylor that bone is less of an acceptable relic in the period, but I argue here in Chapter 5 that there is still a desire for skeletal remains, and that ivory replaces human bone in mourning objects.

40. Laver's billhead is part of the ephemera collection at the British Museum.

41. Margaret K. Powell and Joseph Roach, eds., "Introduction," in *A Cultural History of Hair in the Age of Enlightenment* (London: Bloomsbury, 2020).

42. Jayne Lewis, "'A Lock of Thy Bright Hair': The Enlightenment's Milton and Our Auratic Material," *Humanities* 4 (2015), 799.

43. Helen Sheumaker, *Love Entwined: The Curious History of Hairwork in America* (Philadelphia: University of Pennsylvania Press, 2007), 9.

44. Christiane Holm, "Sentimental Cuts: Eighteenth-Century Mourning Jewelry with Hair," *Eighteenth-Century Studies* 38, 1 (2004), 140.

45. Fennetaux, "Fashioning Death/Gendering Sentiment," 32.

46. Elizabeth Justice, *Amelia, or, The Distress'd wife: a history founded on real circumstances. By a private gentlewoman* (London, 1751), 230.

47. Helen Craik, *Adelaide de Narbonne, with Memoirs of Charlotte de Cordet [. . .]* (London, 1800), 4:67.

48. See Deborah Lutz, *The Brontë Cabinet: Three Lives in Nine Objects* (New York: W. W. Norton, 2015), for her analysis of the hair bracelet intertwined with Anne's and Emily Brontë's hair. The popularity of hair bracelets peaked in the Victorian period, therefore they circulate more frequently in the Victorian novel. For example, in Wilkie Collins's *Hide and Seek* (1854), the omniscient narrator declares that a hair bracelet is "in England one of the commonest ornaments of woman's wear." In *Our Mutual Friend* (1864), Dickens has a character pronounce that watch fobs made of hair were a mark of a middle-class man's respectability. As the extreme sentimentality of the Romantic and Victorian periods faded, "hair from a deceased body was no longer seen as being the delicate remembrance of a life now passed but was looked at as being unhealthy and something that might harbor disease." See Lutz, *Relics of Death*, 134, as well as D. Tulla Lightfoot, *The Culture and Art of Death in 19th Century America* (Jefferson, NC: McFarland, 2019), 146.

49. For more on classical and neoclassical mourning imagery, see Kate A. Beats, "Enduring Grief: Images of the Mourning Woman from the Ancient Classical World to Eighteenth-Century Britain," in *The Materiality of Mourning: Cross-Disciplinary Perspectives*, ed. Zahra Newby and Ruth E. Toulson (New York: Routledge, 2018), 148–71.

50. Ann Radcliffe, *The Romance of the Forest* (Oxford: Oxford University Press, 2009), 272–73.

51. Charlotte Palmer, *It is, and it is not a novel* (London, 1792), 296.

52. Goethe's novel was readily refashioned by the English: an ever-expanding panoply of suicidal heroes appears in Herbert Croft's *Love and Madness; or, A Story Too True* (1780) and Jacobin novels such as Charlotte Smith's *Desmond* (1792). For an excellent study of suicide, sensibility, and the novel, see Kelly McGuire, *Dying to Be English: Suicide Narratives and National Identity, 1721–1814* (New York: Routledge, 2012).

53. Though the fictional examples I discuss include locks of hair from known donors, the ambiguity and anonymity of hair in miniatures and commemorative jewelry must also be mentioned. Christiane Holm notes this troubling discrepancy between commemoration and commodity: "In baroque mourning jewels hair is mostly presented in a simple single lock. During the eighteenth century new techniques were developed to convert the fine material into artificial forms. These culminated in the hair-industry of the nineteenth century, the significant forms of which have been characterized as an increased hiding or disguising of the material's bodily origin" (140).

54. Georgiana Spencer Cavendish, Duchess of Devonshire (1757–1806), *The Sylph* (Evanston, IL: Northwestern University Press, 2007), 129.

55. Jane Austen, *Sense and Sensibility* (Oxford: Oxford University Press, 2008), 74. For a helpful discussion of miniatures and hairwork in Austen, see Kristen Miller Zohn's "Tokens of Imperfect Affection: Portrait Miniatures and Hairwork in *Sense and Sensibility*," *Persuasions On-Line* 32, no. 1 (Winter 2011), http://www.jasna.org/persuasions/on-line/vol32no1/zohn.html?.

56. For an insightful discussion of Belinda's lock of hair, and its connection to *Clarissa*, see Sean Silver, *The Mind Is a Collection: Case Studies in Eighteenth-Century Thought* (Philadelphia: University of Pennsylvania Press, 2015).

57. Jeanenne Bell, *Collector's Encyclopedia of Hairwork Jewelry: Identification & Values* (Paducah, KY: Collector Books, 1998), 21. The Victoria and Albert Museum collection houses numerous delicate examples of eighteenth-century hair jewelry, both love tokens and memorial objects.

58. Near the end of the novel, Lucy writes to Edward to announce she is engaged to his brother, offering to return his portrait and stating he can keep the ring.

59. Charlotte Smith, *Celestina* (Ontario: Broadview, 2004), 382.

60. Miss Milner marries Dorriforth when he is Catholic, and then in a second ceremony when he converts to Protestantism and gives up his rites as a Catholic priest.

61. Elizabeth Inchbald, *A Simple Story* (Oxford: Oxford University Press, 1998), 193.

62. Frank, *Love and Loss*, 147.

63. Marcia Pointon, "Materializing Mourning: Hair, Jewellery and the Body in *Material Memories: Design and Evocation* (Oxford: Oxford University Press, 1999), 56.

64. Pointon, "Materializing Mourning," 56.

65. Pointon, "Materializing Mourning," 56.

66. Keats's "Lines on Seeing a Lock of Milton's Hair" (1818) certainly comes to mind here, as Milton (as with many others) is poetically elevated to saintly realms, and as Keats's legend will posthumously experience. Notoriously, Robert Browning supposedly kept a lock of Milton's hair, along with a lock of Elizabeth Barrett's.

67. Philip Neve, *Narrative of the Disinterment of Milton's Coffin* (London, 1790), 23. William Cowper's poem "On the Late Indecent Liberties Taken with The Remains of the Great Milton, Anno 1790" (1803) details the breaking up of Milton's corpse by the church officials and viewing public. See Lorna Clymer's "Cromwell's Head and Milton's Hair: Corpse Theory in Spectacular Bodies of the Interregnum," *The Eighteenth Century* 40, no. 2 (1999): 91–112.

68. B. B., "Milton's Rib-bone," *Notes & Queries*, April 17, 1852, 369.

69. *Keats's 23 January 1818 Letter to Benjamin Bailey*. For helpful background, see Jayne Lewis, The Keats Letters Project, "Keats's Good Hair Day" (2018), https://keatslettersproject.com/correspondence/keatss-good-hair-day/. Lewis remarks, "Hunt was in fact an inveterate collector of the locks of the famous; Milton's was part of a larger collection that, at final tally, would include the locks of (among others) Swift, Johnson, Coleridge, Wordsworth, Maria Edgeworth, and Percy Shelley—a hirsute anthology of English literary history that Hunt eventually annotated with the tresses of Napoleon, George Washington and Lucrezia Borgia. You can visit Hunt's collection today in the Ransom Library at the University of Texas."

70. See Richard Oram, "The Locks of Ages: The Leigh Hunt Hair Collection," *Harry Ransom Center Magazine* (January 13, 2011), https://sites.utexas.edu/ransomcentermagazine/2011/01/13/locks-of-ages-the-leigh-hunt-hair-collection/.

71. Charlotte Smith, *Ethelinde* (Toronto: Broadview, 2016), 70. Typical mourning called for a plain style with no ruffles, no shiny buttons or buckles, but there are numerous examples housed in the Victoria and Albert Museum that are set with brilliants and hair (most likely from the third period of mourning and for half mourning). Later, when Queen Victoria mourned her husband Prince Albert's death in 1861, she started the trend of wearing mourning buttons made of jet.

72. In "Fashioning Death/Gendering Sentiment," Fennetaux effectively argues for an increasingly feminine excess in terms of Victorian mourning jewelry.

73. Maxine Berg, *Luxury and Pleasure in Eighteenth-Century Britain* (Oxford: Oxford University Press, 2007), 3.

74. I wish to thank Alice Minter, curator, the Rosalinde and Arthur Gilbert Collection, and Richard Edgcumbe, senior curator of the Victoria and Albert Museum, for their suggestion of this item. I am also indebted to Rachel Church, curator of sculpture, metalwork, ceramics, and glass at the Victoria and Albert Museum, for her deep knowledge of period mourning rings.

75. Quoted in *The London and Dublin Magazine: Or, Gentleman's Monthly Intelligencer* (London, 1762). Volume 24 of the 1762 *The Scot's Magazine* describes another related toothpick case: "And likewise a toothpick-case, curiously ornamented with silver, made of the piece of the oak which K. Charles II. cut from the tree while secreted there from the pursuit of his enemies; on the top is engraved a crown, and the words Roy AL OAk" (393).

CHAPTER 5

1. For a helpful cultural history of ivory, see John Frederick Walker, *Ivory's Ghosts: The White Gold of History and the Fate of Elephants* (New York: Grove Atlantic, 2010), 74. Ivory is not always associated with the beauty of the female body. In John Cleland's *Memoirs of a Woman of Pleasure* (1749), Fanny Hill is aroused by the "maypole" of her male lover, exclaiming "such a length, such a breadth of animated ivory!").

2. Eliza Haywood, *The History of Miss Betsy Thoughtless* (Oxford: Oxford University Press, 1997), 392.

3. Henry Fielding, *Amelia* (London: Penguin, 1987), 102.

4. Jakub Lipski, *Painting the Novel: Pictorial Discourse in Eighteenth-Century English Fiction* (New York: Routledge, 2017), 50.

5. Orest Ranum, "The Refuges of Intimacy," *A History of Private Life*, vol. 3, *Passions of the Renaissance*, ed. Roger Chartier (Cambridge, UK: Cambridge University Press, 1989), 207–63.

6. Marcia Pointon, *Hanging the Head: Portraiture and Social Formation in Eighteenth-Century England* (New Haven, CT: Yale University Press, 1993), 53.

7. Some notable studies of the English miniature include Graham Reynolds, *English Portrait Miniatures* (Cambridge, UK: Cambridge University Press, 1988); Joe Bray, *The Portrait in Fiction of the Romantic Period* (New York: Routledge, 2016); Marcia Pointon, *Hanging the Head*; and Katherine Coombs, *The Portrait Miniature in England* (London: Victoria and Albert Museum, 1998).

8. Susan Stewart, *On Longing: Narratives of the Miniature, the Gigantic, the Souvenir, the Collection* (Durham, NC: Duke University Press, 1993), 69.

9. My use of "mourning miniature" connotes portraits commissioned for recipients specifically as a mourning memento as well as portraits used as mourning talismans but not originally intended for that specific purpose.

10. Samuel Richardson, *The History of Sir Charles Grandison* (Oxford: Oxford University Press, 1986), letter 20, 2:368–69.

11. Elizabeth Griffith, *The Delicate Distress* (Lexington: University Press of Kentucky, 1997), 78.

12. Griffith's novel includes numerous examples of miniatures. One scene is especially dramatic: Lord Woodville "took out the marchioness's picture, and hung it round my neck, as a kind of talisman, against that remorse which I must certainly feel, for abandoning my wife" (214). He in fact is gravely injured from a fall from his horse, and the miniature reveals to all his infidelity. The novel's final lines describe Lord Seymour being presented with two male miniatures: that of Charlotte's brother Captain Beaumont (who is living) and Charlotte's father, who has died.

13. Ann Radcliffe's *A Sicilian Romance* (Oxford: Oxford University Press, 1998), 27. Later, banditti attack and steal a miniature from Hippolitus: "They were proceeding to take from him a miniature picture, which was fastened round his neck, and had been hitherto concealed in his bosom," 161. Differing from my interpretation, Bray reads the role of the miniature and likeness as confusing identity in period novels.

14. Ann Radcliffe, *The Mysteries of Udolpho* (Oxford: Oxford University Press, 1998), 26.

15. See Robin Jaffee Frank, *Love and Loss* (New Haven, CT: Yale University Art Gallery) for a discussion of the gendered wearing of miniatures, 15.

16. Melinda Rabb, *Miniature and the English Imagination: Literature, Cognition, and Small-Scale Culture, 1650–1765* (Cambridge, UK: Cambridge University Press, 2019), 92n10.

17. In Radcliffe's *The Italian* (1797), the orphan Ellena experiences a similar birthright plot that is hinged on her resemblance to her father, also depicted in a mourning miniature. Joe Bray discusses the role of miniature portraits (not specifically their mourning roles) in Radcliffe's fiction. See *The Portrait in Fiction of the Romantic Period* (New York: Routledge, 2016).

18. There are few novels that center their birthright plots on mourning rings. One notable example is Elizabeth Griffin's *The friends; or, the contrast between virtue and vice. A tale. Designed for the improvement of youth. By Elizabeth Griffin, Author of the Selector, and Moral Amusements, &c. &c.* (London, 1799). Here, a mourning ring that is the property of the deceased Mrs. Curtis "was carefully preserved" by Charles "as there was a probability that it might one day tend to elucidate the family from whence Charles Curtis derived his origin" (63). Indeed, the Curtis

family maid gives Charles the mourning ring, telling him "to let nothing induce me to part with it, as it might be the means, one time or other, of restoring me to those who had the best right to take care of me" (54).

19. Charlotte Smith, *Emmeline, The Orphan of the Castle* (Ontario: Broadview, 2003), 70.

20. See Frank's excellent analysis of P. R. Vallée's *Harriet Mackie (The Dead Bride)* (c. 1804), which notes the practice of miniatures taken from the dead body and explains how rare it would be for a miniaturist to depict the dead with closed eyes. *Love and Loss*, 141–53. The genre of the deathbed portrait was popular in Reformation Europe and Britain (examples include that of Venetia Stanley [*Lady Digby, on her Deathbed*, c. 1633], Sir Thomas Aston, and the Saltonstall Family). In the seventeenth century, John Donne (d. 1631) had a portrait taken as if he were dead so the cadaver tomb that he commissioned for St. Paul's Cathedral would be accurate. Despite the supposed frequency of deathbed portraits commissioned in the period, existing English miniatures are scant. For example, the extensive catalog of seventeenth- and eighteenth-century miniatures for the Fitzwilliam Museum does not include a deathbed portrait.

21. Kate A. Beats, "Enduring Grief," in *The Materiality of Mourning: Cross-Disciplinary Perspectives*, ed. Zahra Newby and Ruth E. Toulson (New York: Routledge, 2018), 163. Beats's discussion of Wedgwood urns is especially informative.

22. See also Katherine Coombs, *The Portrait Miniature in England* (London: V&A Publications, 2005).

23. In the Victorian period, machines were patented that produced long, exquisitely thin sheets of ivory. This allowed the production not only of ivory veneers for furniture but of the thin plates need for the painting of miniatures. See John Frederick Walker's fascinating study, *Ivory's Ghosts: The White Gold of History and the Fate of Elephants* (New York: Grove Atlantic, 2010).

24. For helpful discussions of mourning, ivory, and bas-reliefs, see Zhara Newby, "The Grottarossa Doll and Her Mistress: Hope and Consolation in a Roman Tomb," as well as Ruth E. Toulson, "Fragments of Bone and Chips of Stone: Materiality and Mourning in Chinese Society," in *The Materiality of Mourning: Cross-Disciplinary Perspectives*, ed. Zahra Newby and Ruth E. Toulson (New York: Routledge, 2018), 77–102 and 191–204.

25. Joseph Addison, *The Spectator*, no. 245 (December 11, 1711).

26. Kamilla Elliott, *Portraiture and British Gothic Fiction: The Rise of Picture Identification, 1764–1835* (Baltimore: Johns Hopkins University Press, 2012), 67.

27. Evans, J, "On the Utility of Paintings," *Scots Magazine* 57 (July 1795): 433–34.

28. Frank remarks, "It was rare for a miniaturist to depict the dead with closed eyes . . . usually the eyes were shown open, so that the memory of the *living* person could be preserved" (141). By the early nineteenth century, memorial miniatures actually include "tear jewelry," sentimental keepsakes typically depicting a close-up of one tearful eye.

29. Anonymous, "Talismans," *Conjuror's Magazine; or, Magical and Physiognomical Mirror* (1791), 183.

30. Richard Cosway, "Margaret Cocks, later Margaret Smith," EMuseum.huntington .org. The Huntington Library, Art Collections, and Botanical Gardens image notes state that "In the present portrait, the framed oval miniature suspended by a chain from the sitter's neck alludes to his more characteristic mode. The miniature also provides one of a number of clues indicating that the painting's thematic subtext is the longing of young Margaret Cocks for an absent loved one, presumably the person represented in the miniature. Fortunately, Cosway has delineated the miniature so precisely that it is possible to identify its subject as Margaret's elder sister, Mary (Cocks) Russell, as represented by John Smart in a miniature signed and dated 1781. Co-

sway set the emotional tone of the portrait by modeling Margaret Cocks's pose on the ancient iconography of Melancholy, which had been widely disseminated in modern Europe by sixteenth-century Northern European engravings. The pose became a cliché of female portraiture in late eighteenth-century Britain, valued as a fashionably neo-classical embodiment of refined sensibilities." http://emuseum.huntington.org/objects/2185/margaret-cocks-later-margaret-smith?ctx=19a346d6-77ff-46f5-bf6a-b3477a3562fc&idx=3.

31. Mary Shelley, *Frankenstein* (New York: W. W. Norton, 2011), 97.

32. Shelley, *Frankenstein*, 49.

33. Beats, "Enduring Grief," 165. While I acknowledge the aesthetic properties of these images, and have argued for a rise of the "beautiful death" in the century, I read images that depict the physical engagement with dead bodies (such as embracing urns) as a desire for corporeal contact and representation, not a fear of the corpse.

34. Frank remarks that "In England the earliest known painting of a miniature being worn by a woman dates from around 1560, during the reign of Queen Elizabeth, and the wearer was herself royalty—Lady Katherine Grey. By the 1570s, miniatures in England had become available to the gentry and those of the 'middling rank' as well." See *Love and Loss*, 15–18.

35. Clara Reeve's *The Old English Baron* (Oxford: Oxford University Press, 2008), 102. I wish to thank Jeremy Chow for suggesting Reeve's novel.

36. Horace Walpole, author of *The Castle of Otranto*, which deeply influenced Reeve's novel, on reading *The Old English Baron*, writes a scathing review in a letter, to include, "This is a *caput mortuum*," that is, a dead head or worthless remains. I find this phrase ironically apt for a plot centered on a murder mystery solved by skeletal remains. The bones that circulate in the novel are anything but "worthless remains." Horace Walpole to William Mason, April 8, 1778.

37. Ann Radcliffe, *The Romance of the Forest* (Oxford: Oxford University Press, 2009), 53–54. Yael Shapira cites an incident from Mrs. Carver's *The Horrors of Oakendale Abbey* (1797), where the heroine Laura is faced with numerous morbid objects, including a "ghastly skeleton" hidden in a trunk: Coming across a large trunk, Laura opens it and is "struck with horror and astonishment, when the skeleton of a human body presented itself to her affrighted view! She gave an involuntary scream, and dropping the lid from her trembling hand, the sound, echoing through the hallow roof, vibrated with terror upon her palpitating heart" (164). Later, Charlotte Dacre will utilize the "skeleton in the chest" trope for more morbid and shocking purposes in *Zofloya* (1806): "With horror and perturbation in his looks, Antonio seemed rapidly advancing towards the chest, and calling aloud for some of his companions to assist him, by their joint efforts they raised the lid.—This was no sooner accomplished than a shout of universal horror prevailed, accompanied by the strongest marks of terror and perturbation. The cause was presently explained.—Forth from the chest they drew the disclosed, half-mouldered skeleton, that once had been Berenza!" (Charlotte Dacre, *Zofloya, or The Moor* [Oxford: Oxford University Press, 1997], 228).

38. As Yael Shapira notes, "the dead body indeed becomes one of the critics' favorite tropes for denouncing the crude stimuli of popular fiction" (*Inventing the Gothic Corpse*, 10). The anonymous author of "Terrorist Novel-Writing" ("The Spirit of the Public Journals," 1797) states, "If a curtain is withdrawn, there is a bleeding body behind it; if a chest is opened it contains a skeleton." Later, Jane Austen will parody this in *Northanger Abbey* (1818), in which the heroine Catherine Morland will excite herself as she reads a climactic moment from *The Mysteries of Udolpho*: "I know it must be a skeleton, I am sure it is Laurentina's skeleton. Oh! I am delighted with the book! I should like to spend my whole life in reading it."

39. Damer is buried in the church in Sundridge where her mother lies buried. See Jonathan David Gross, *The Life of Anne Damer: Portrait of a Regency Artist* (Lanham, MD: Lexington

Books, 2013). Lord Byron, a notorious dog lover, left orders that he be buried in the family vault with his dog: "The body of Lord B to be buried in the vault of the garden of Newstead, without any ceremony or burial service whatever, or an inscription, save his name and age. His dog not to be moved from the said vault." These directions were not executed, but he did erect a monument to his dog, Bosun. See Gross for images of the monument and an illustration of Bosun.

40. Anita Guerrini, "Anatomists and Entrepreneurs in Early Eighteenth-Century London," *Journal of the History of Medicine and Allied Sciences* 59 (2004): 229

41. Guerrini notes that the Company of Barber-Surgeons reprimanded Cheselden in 1714 for procuring "dead bodies of malefactors from the place of execution" and dissecting them in his house. Guerrini, "Anatomists and Entrepreneurs," 238. Londa Schiebinger examines Cheselden's illustrations of the female skeletons, noting that previous to Cheselden, the male skeleton represented the general form of the human body's foundation. See "Skeletons in the Closet: The First Illustrations of the Female Skeleton in Eighteenth-Century Anatomy," *Representations* 14 (1986): 42–81. Of interest to my discussion of the senses in Chapter 1, in 1728 Cheselden also published an account of cataract surgery on a blind person, which resulted in the partial regaining of sight. See Chapter 1, note 94, in this volume.

42. Simon Chaplin, "John Hunter and the 'Museum Oeconomy,' 1750–1800," (PhD diss., University of London, 2009), 51–54. See the Wellcome Institute site for reference (Chaplin is also the director of culture and society), https://wellcomecollection.org.

43. After the passing of the 1752 Murder Act, which added the infamy of a public autopsy to the punishment for murder, crowds attended public dissections at Surgeons' Hall. Paying spectators of both sexes visited anatomical museums to view preserved body parts and wax models of the body's interior.

44. Shapira, *Inventing the Gothic Corpse*, 200. For a helpful discussion of phantasmagoric technologies, see David Jones, *Gothic Machine: Textualities, Pre-Cinematic Media and Film in Popular Visual Culture, 1670–1910* (Wales: Wales University Press, 2011).

45. The Royal College of Surgeons had a legal right to the corpses of condemned criminals and articulated many following dissection. For a brilliant analysis of Wild, Defoe's *True and Genuine Account of the Life and Actions of the Late Jonathan Wild* (1725), and ownership and property, see Sean Silver's Case 6 study in *The Mind Is a Collection: Case Studies in Eighteenth-Century Thought* (Philadelphia: University of Pennsylvania Press, 2015).

46. Guerrini, "Anatomists and Entrepreneurs," 230.

47. For a fascinating history of the medicinal use of "mummy" in the period (as well as human skull), see Richard Sugg, *Mummies, Cannibals, and Vampires: The History of Corpse Medicine from the Renaissance to the Victorians* (New York: Routledge, 2011).

48. Helen Deutsch, "'Alas Poor Yorick!': Jonathan Swift, Madness and Fashionable Science," in *Disease and Death in Eighteenth-Century Literature and Culture: Fashioning the Unfashionable*, ed. Allan Ingram and Leigh Wetherall Dickson (London: Palgrave Macmillan), 225–45. Also notable is that according to Jayne Lewis, "Swift's hair was trimmed on his deathbed and distributed among his friends, who perhaps identified it with the hair that apparently made Gulliver so vulnerable to the Lilliputians (they use it to tie him down)." See https://keatslettersproject.com/correspondence/keatss-good-hair-day/.

49. *National Quarterly Review* 20 (1870) records that "His grave was opened by order of Captain James Shuttleworth, and a great thigh-bone was found in it, which measured thirty-two inches in length. This relic was re-interred" (107). The writer goes on to note that "These doings, even if they *prove* nothing as to the fact of Little John's existence, show how strong is the belief in it at the present day."

50. See Bess Lovejoy, *Rest in Pieces: The Curious Fates of Famous Corpses* (New York: Simon & Schuster, 2013), for further discussion of the skull fragments taken from Shelley's funeral pyre July 8, 1822. The website *Atlas Obscura* includes Lovejoy's helpful explanation of how Pforzheimer received the fragments, as well as photographs of the skull remnants and attestations. Bess Lovejoy, "How Did Bits of Percy Shelley's Skull End Up in the New York Public Library?" *Atlas Obscura*, July 8, 2013, https://www.atlasobscura.com/articles/shelley-skull-fragments-at-nypl. There are numerous cases of eighteenth-century celebrity skulls. One notable legend is that of Edward Teach, the pirate Blackbeard. When British naval commander Lieutenant Robert Maynard ambushed Blackbeard's ship on November 22, 1718, he shot Teach, beheaded his corpse, and dangled his head from the front of his ship. When he returned to Virginia, Maynard gave the skull to the lieutenant governor of Virginia, Alexander Spotwood. Legend has it that the skull fell into the hands of a Williamsburg tavern owner, who had the skull lined with silver and "dared his patrons to drink from it." The silver skull is now part of the Peabody Essex Museum, Salem, Massachusetts (and is not displayed). For details and period illustrations, see Dolly Stolze's fascinating website, Strange Remains (2013–2018), https://strangeremains.com/2018/04/02/the-legend-of-blackbeards-silver-plated-skull/.

Notoriously, George Gordon, Lord Byron had a skull cup made from a monk's skull unearthed by a gardener at Newstead Abbey. His humorous poem "Lines Inscribed Upon a Cup Formed from a Skull" (1808) includes the question, "our brains are gone, / What nobler substitute than wine?"

51. See Sean Silver's analysis of body stones, the evocative account of Samuel Pepys's lithotomy and revered bladder stones, as well as the Hunterian Museum's collection of stones and their receptacles in *The Mind Is a Collection*.

52. Richardson and Hurwitz, *British Medical Journal*, 197. The board of trustees of the Hunterian Museum is currently considering giving Charles Byrne a proper burial. See Hannah Devlin, "'Irish Giant' May Finally Get Respectful Burial After 200 Years on Display," *The Guardian*, June 22, 2018, https://www.theguardian.com/science/2018/jun/22/irish-giant-may-finally-get-respectful-burial-after-200-years-on-display. Hilary Mantel's novel *The Giant, O'Brien* (1998) dramatizes John Hunter's directions to grave robbers for stealing Byrne's corpse. The Royal College's Hunterian Museum also displays the skeleton of famed "thief-taker" Jonathan Wild. Convicted and hung at Tyburn on May 24, 1725, for his criminal activities (tickets were sold to the public), Wild was secretly buried in St. Pancras' churchyard, but his grave was quickly spotted by body snatchers and his corpse was sold to a surgeon for dissection and articulation. The skeleton was received in 1847 by the Royal College of Surgeons, where it is still displayed.

53. See Richard Burt, "DIE-JESTING stURNe's BURIALLs: Publication, Plagiarism, Pseudonymity, Pseudography, Cenography, Palimpsestuosity, Posthumography, and the Propriety or Pathos of Posterity," in *Shakespeare's Hamlet in an Era of Textual Exhaustion*, ed. Allison Kellar Lenhardt, Sonya Loftis, and Lisa Ulevich (New York: Routledge, 2018), 199–243; and W. G. Day, "The Skull," *The Shandean* 10 (1998): 45–79. My special thanks to Professor Peter de Voogd of the International Laurence Sterne Foundation for generously sharing his photograph of Sterne's skull.

54. *Public Advertiser* (1769).

55. Day, "The Skull," 53. Numerous sources state that those present at the dissection immediately recognized Sterne's face. Sterne's features were noted to be distinct, and he frequently alluded to his own appearance in conversation, letters, and other writings.

56. Ian Campbell Ross, *Laurence Sterne: A Life* (Oxford: Oxford University Press, 2001), 19.

AFTERWORD

1. See Ruth Richardson, *Death, Dissection and the Destitute* (New York: Routledge, 1987) for a detailed account of the culture of mourning, burial reform, the Anatomy Act, and their effects on the poor. For other Victorian death studies, see John Morley, *Death. Heaven and the Victorians* (London: Studio Vista, 1971); James Curl, *The Victorian Celebration of Death* (Detroit: Partridge Press, 1972); John Reed, *Victorian Conventions* (Columbus: Ohio University Press, 1975); Michael Wheeler, *Death and the Future Life in Victorian Literature and Theology* (Cambridge, UK: Cambridge University Press, 1990); Esther Schor, *Bearing the Dead: The British Culture of Mourning from the Enlightenment to Victoria* (Princeton, NJ: Princeton University Press, 1994); Pat Jalland, *Death in the Victorian Family* (Oxford: Oxford University Press, 1996).

2. Wheeler, introduction, *Death and the Future Life in Victorian Literature and Theology*, xiii.

3. Richardson, *Death, Dissection and the Destitute*, 7. Richardson provides a detailed discussion of the Anatomy Act, burial reform, and the nineteenth-century understanding of the body-soul dichotomy.

4. Jolene Zigarovich, *Writing Death and Absence in the Victorian Novel: Engraved Narratives* (New York: Palgrave, 2012), 4. For further discussion of the rise of death in literary culture, see Zigarovich's introduction, 1–22, and Zigarovich, "Death, Literature, and the Victorian Era," in *The Routledge Companion to Death and Literature*, ed. W. Michelle Wang, Daniel K. Jernigan, and Neil Murphy (New York: Routledge, 2021), 288–97.

5. David Cannadine, "War and Death, Grief and Mourning in Modern Britain," in *Mirrors of Mortality: Studies in the Social History of Death* (New York: St. Martin's, 1981), 191. Cannadine recognizes further that "the whole obsessive paraphernalia of mourning," such as lockets, earrings, and so forth, "were more a cause of financial anxiety to the bereaved than a source of emotional solace."

6. Pat Jalland, *Death in the Victorian Family* (Oxford: Oxford University Press, 1996), 305, 200–203. Jalland also notes the decline of the ostentatious funeral among the middle classes in the second half of the century.

7. Claire Wood, *Dickens and the Business of Death* (Cambridge, UK: Cambridge University Press, 2015), 11. See chapter 1 of Wood's volume for a thorough examination of Victorian middle-class death-related consumption.

8. John Kucich, *Excess and Restraint in the Novels of Charles Dickens* (Athens: University of Georgia Press, 1981), 177.

9. Deborah Lutz, *Relics of Death in Victorian Literature and Culture* (Cambridge, UK: Cambridge University Press, 2015), 167.

10. Lutz, *Relics of Death*.

BIBLIOGRAPHY

Abernethy, John. "Observations on the Foramina Thebesii of the Heart." *Philosophical Transactions* 88 (1798): 103–9.

Adams, Ellen. "Defining and Displaying the Human Body: Collectors and Classics During the British Enlightenment." *Hermathena*, no. 187 (Winter 2009): 65–97.

Addison, Joseph. "Pleasures of Imagination." *The Spectator*, June 21, 1712.

———. *The Spectator*, December 11, 1711.

Alberti, Samuel. *Morbid Curiosities: Medical Museums in Nineteenth-Century Britain*. Oxford: Oxford University Press, 2011.

———. "Wax Bodies: Art and Anatomy in Victorian Medical Museums." *Museum History Journal* 2, no. 1 (2009): 7–36.

Almond, Philip C. *Heaven and Hell in Enlightenment England*. Cambridge, UK: Cambridge University Press, 1994.

Altick, Richard D. *The Shows of London*. Cambridge, MA: Belknap Press, 1978.

Anonymous. *The Death, Dissection, Will, and Funeral Procession, of Mrs. Regency Revised, Corrected, Altered, Added to, and Considerably Amended, by the Author*. London: Printed at the Logographic Press, 1789.

———. *Extraordinary, and Unparallel'd Curiosities, Composed of Wax, (so near the Life, That They Cannot Be Excelled by Art) in Seven Several Scenes*. London, 1664.

———. "Some Account of a Body Lately Found in Uncommon Preservation [. . .]." *Philosophical Transactions of the Royal Society of London* 62 (1772): 465–68.

———. "Talismans." *Conjuror's Magazine; or, Magical and Physiognomical Mirror*, 1791.

Appleby, John H. "Human Curiosities and the Royal Society, 1699–1751." *Notes and Records of the Royal Society of London* 50, no. 1 (1996): 13–27.

Ariès, Philippe. *The Hour of Our Death*. Translated by Helen Weaver. New York: Knopf, 1981.

———. *Western Attitudes Toward Death: From the Middle Ages to the Present*. Translated by Patricia M. Ranum. Johns Hopkins Symposia in Comparative History. Baltimore: Johns Hopkins University Press, 1974.

Armstrong, Nancy. *Desire and Domestic Fiction: A Political History of the Novel*. Oxford: Oxford University Press, 1987.

———. *How Novels Think: The Limits of British Individualism from 1719–1900*. New York: Columbia University Press, 2005.

Aronson, Julie, and Marjorie E. Wieseman. *Perfect Likeness: European and American Portrait Miniatures from the Cincinnati Art Museum*. New Haven, CT: Yale University Press, 2006.

Aubin, Robert A. "Behind Steele's Satire on Undertakers." *PMLA* 64, no. 5 (1949): 1008–26.

Austen, Jane. *Sense and Sensibility*. Oxford: Oxford University Press, 2008.

Backscheider, Paula R. *Elizabeth Singer Rowe and the Development of the English Novel.* Baltimore: Johns Hopkins University Press, 2013.

Baillie, Matthew. "On the Embalming of Dead Bodies." *Transactions of a Society for the Improvement of Medical and Chirurgical Knowledge* 3 (1812): 7–23.

Balguy, Charles. "An Account of the Dead Bodies of a Man and Woman, Which Were Preserved 49 Years in the Moors in Derbyshire [. . .]." *Philosophical Transactions* 38 (1733): 413–15.

Bancroft, Francis. *A True Copy of the Remarkable Last Will and Testament of Mr. Francis Bancroft, Citizen and Draper of London.* London: Printed for J. Peele, 1728.

Barker-Benfield, G. J. *The Culture of Sensibility: Sex and Society in Eighteenth-Century Britain.* Chicago: University of Chicago Press, 1996.

Barnett, Richard. "Lost Wax: Medicine and Spectacle in Enlightenment London." *Lancet* 372, no. 9636 (2008): 366–67.

Barr, Rebecca Anne, Sylvie Kleiman-Lafon, and Sophie Vasset, eds. *Bellies, Bowels and Entrails in the Eighteenth Century.* Manchester, UK: Manchester University Press, 2019.

Bates, A. W. "'Indecent and Demoralising Representations': Public Anatomy Museums in Mid-Victorian England." *Medical History* 52, no. 1 (2008): 1–22.

Baxandall, Michael. *Shadows and Enlightenment.* New Haven, CT: Yale University Press, 1995.

Beats, Kate A. "Enduring Grief: Images of the Mourning Woman from the Ancient Classical World to Eighteenth-Century Britain." In *The Materiality of Mourning: Cross-Disciplinary Perspectives,* edited by Zahra Newby and Ruth E. Toulson. New York: Routledge, 2018.

Becker, Lucinda M. *Death and the Early Modern Englishwoman.* Aldershot, UK: Ashgate, 2003.

Bell, Jeanenne. *Collector's Encyclopedia of Hairwork Jewelry: Identification & Values.* Paducah, KY: Collector Books, 1998.

Bell, Michael. *Sentimentalism, Ethics, and the Culture of Feeling.* London: Palgrave Macmillan, 2000.

Berg, Maxine. *Luxury and Pleasure in Eighteenth-Century Britain.* Oxford: Oxford University Press, 2005.

Berkeley, George. "An Essay Toward a New Theory of Vision." In *Philosophical Works, Including the Works on Vision,* edited by Michael R. Ayers. London: J. M. Dent, 1975.

Berridge, Kate. *Madame Tussaud: A Life in Wax.* New York: Harper Perennial, 2006.

Bland, Olivia. *The Royal Way of Death.* London: Constable, 1986.

Bloom, Michelle E. *Waxworks: A Cultural Obsession.* Minneapolis: University of Minnesota Press, 2003.

Bloxam, Matthew Holbeche. *A Glimpse at the Monumental Architecture and Sculpture of Great Britain [. . .].* Whitefish, MT: Kessinger, 2007.

Blumenbach, John Frederick. "Observations on Some Egyptian Mummies Opened in London." *Philosophical Transactions of the Royal Society of London* 84 (1794): 177–95.

Bowden, John. *The Grave, the House Appointed for All Living [. . .].* ECCO. Bristol: Printed for Thomas Cadell, 1749.

Bradford, Charles. *Heart Burial.* London: Allen and Unwin, 1933.

Bray, Joe. *The Portrait in Fiction of the Romantic Period.* Abingdon, UK: Routledge, 2016.

Brett, Mary. *Fashionable Mourning Jewelry, Clothing & Customs.* Atglen, PA: Schiffer, 2006.

Brier, Bob. *The Encyclopedia of Mummies.* New York: Facts on File, 1998.

Bronfen, Elisabeth. *Over Her Dead Body: Death, Femininity, and the Aesthetic.* New York: Routledge, 1992.

Brown, Stewart J., and Timothy Tackett, eds. *The Cambridge History of Christianity.* Vol. 7, *Enlightenment, Reawakening and Revolution, 1660–1815.* Cambridge, UK: Cambridge University Press, 2006.

Browne, Thomas. *Hydriotaphia, Urne-Buriall, or, a Discourse of the Sepulchrall Urnes Lately Found in Norfolk [. . .].* London: 1658.

Brownrig, Ralph. *Sixty Five Sermons.* 2 vols. London: Printed by Tho. Roycroft for John Martyn, 1674.

Bryson, Rowan Rose. "The Senses in Literature: Pleasures of Imagining in Poetry and Prose." In *A Cultural History of the Senses in the Age of Enlightenment.* London: Bloomsbury, 2014.

Buck, Anne. *Dress in Eighteenth-Century England.* New York: Holmes & Meier, 1979

Burke, Edmund. *A Philosophical Enquiry into the Origin of Our Ideas of the Sublime and Beautiful.* Project Gutenberg, 2021. https://www.gutenberg.org/files/15043/15043-h/15043-h.htm.

———. *The Works of the Right Honourable Edmund Burke.* Vol. 1. Project Gutenberg, 2005. http://www.gutenberg.org/files/15043/15043-h/15043-h.htm.

Burney, Fanny. *Cecilia, or, Memoirs of an Heiress.* Edited by Peter Sabor and Margaret Anne Doody. Oxford World's Classics. Oxford: Oxford University Press, 1988.

———. *Evelina, or the History of a Young Lady's Entrance into the World.* Oxford: Oxford University Press, 2002.

Burt, Richard. "DIE-JESTING StURNe's BURIALLs: Publication, Plagiarism, Pseudonymity, Pseudography, Cenography, Palimpsestuosity, Posthumography, and the Propriety or Pathos of Posterity." In *Shakespeare's Hamlet in an Era of Textual Exhaustion,* edited by Sonya Loftis and Lisa Uleich, 199–243. New York: Routledge, 2018.

Bury, Shirley. *An Introduction to Sentimental Jewellery.* V & A Introductions to the Decorative Arts. London: H.M.S.O., 1985.

Castle, Terry. *The Female Thermometer: Eighteenth-Century Culture and the Invention of the Uncanny.* New York: Oxford University Press, 1995.

Catalogue of the Contents of the Museum of the Royal College of Surgeons in London. Vol. 9, n.d.

Cavendish, Georgiana Spencer, Duchess of Devonshire. *The Sylph.* Dublin, 1784.

———. *The Sylph.* Edited by Jonathan David Gross. Evanston, IL: Northwestern University Press, 2007.

Chartier, Roger, and Arthur Goldhammer, eds. *Passions of the Renaissance.* A History of Private Life. Cambridge, MA: Belknap Press of Harvard University Press, 1989.

Cheselden, William. "An Account of Some Observations Made by a Young Gentleman, Who Was Born Blind, or Lost His Sight so Early, That He Had No Remembrance of Ever Having Seen, and Was Couch'd between 13 and 14 Years of Age." *Philosophical Transactions* 402 (1728): 447–50.

Church, Rachel. *Rings.* New York: Thames & Hudson, 2017.

Clarkson, Leslie A. *Death, Disease, and Famine in Pre-Industrial England.* New York: St. Martin's Press, 1976.

Classen, Constance. *The Book of Touch.* Oxford: Berg, 2005.

———. *A Cultural History of the Senses.* 6 vols. London: Bloomsbury Academic, 2014.

———. *The Deepest Sense: A Cultural History of Touch.* Studies in Sensory History. Urbana: University of Illinois Press, 2012.

Cleland, John. *Memoirs of a Woman of Pleasure.* Oxford World's Classics. Oxford: Oxford University Press, 2008.

Clymer, Lorna. "Cromwell's Head and Milton's Hair: Corpse Theory in Spectacular Bodies of the Interregnum." *The Eighteenth Century* 40, no. 2 (1999): 91–112.

Collignon, Charles. "XXXIII. Some Account of a Body Lately Found in Uncommon Preservation, under the Ruins of the Abbey, at St. Edmund's-Bury, Suffolk; with Some Reflections upon the Subject." *Philosophical Transactions of the Royal Society of London* 62 (1772): 465–68. https://doi.org/10.1098/rstl.1772.0034.

Collingnon, Charles. "Some Account of a Body Lately Found in Uncommon Preservation, under the Ruins of the Abbey, at St. Edmund's-Bury, Suffolk [. . .]." *Philisophical Transactions*, January 1, 1772.

Condillac, Étienne Bonnot de. *Treatise on Sensations (Traité Des Sensations)*. Stanford Encylopedia of Philosophy, 2017. https://plato.stanford.edu/entries/condillac/.

Conway, Alison, and Mary Helen McMurran, eds. *Mind, Body, Motion, Matter: Eighteenth-Century British and French Literary Perspectives*. Toronto: University of Toronto Press, 2016.

Coolidge, Susan, ed. *The Diary and Letters of Frances Burney, Madame d'Arblay*. Vol. 2. Boston: Roberts Brothers, 1880.

Coombs, Katherine, and Victoria and Albert Museum. *The Portrait Miniature in England*. London: V & A Publications, 2005.

Cope, Kevin L., and Robert C. Leitz, eds. *Sensational Centuries: Essays on the Enhancement of Sense Experience in the Seventeenth, Eighteenth, and Nineteenth Centuries*. New York: AMS Press, 2013.

Cosway, Richard. *Margaret Cocks, Later Margaret Smith*. n.d. EMuseum.huntington.org.

Cox, Margaret. *Life and Death in Spitalfields, 1700 to 1850*. London: Council for British Archaeology, 1966.

Craik, Helen. *Adelaide de Narbonne, with Memoirs of Charlotte de Cordet [. . .]*. Eighteenth Century Collections Online. London: Printed at the Minerva-Press, for William Lane, Leadenhall-Street, 1800.

Cranch, John. *Narrative Relating to the Real Embalmed Head of Oliver Cromwell*. London, 1799.

Craske, Matthew. "Entombed Like an Egyptian: An Eighteenth-Century Surgeon's Extravagant Mausoleum to Preserve His Mortal Remains." *Church Monuments* 15 (2000).

———. *The Silent Rhetoric of the Body: A History of Monumental Sculpture and Commemorative Art in England, 1720–1770*. New Haven, CT: Yale University Press, 2008.

———. "'Unwholesome' and 'Pornographic': A Reassessment of the Place of Rackstrow's Museum in the Story of Eighteenth-Century Anatomical Collection and Exhibition." *Journal of the History of Collections* 23, no. 1 (2011): 75–99.

Cressy, David. *Birth, Marriage, and Death: Ritual, Religion, and the Life-Cycle in Tudor and Stuart England*. Oxford: Oxford University Press, 1999.

Croft, Sir Herbert. *Love and Madness: A Story Too True: In a Series of Letters [. . .]*. 4th ed. Eighteenth Century Collections Online. London, 1780.

Csengei, Ildiko. *Sympathy, Sensibility and the Literature of Feeling in the Eighteenth Century*. London: Palgrave Macmillan, 2012.

Cunningham, Andrew. *The Anatomist Anatomis'd: An Experimental Discipline in Enlightenment Europe*. The History of Medicine in Context. London: Routledge, 2016.

Cunnington, Phillis, and Catherine Lucas. *Costumes for Births, Marriages & Deaths*. London: A. and C. Black, 1972.

Curl, James Stevens. *Egyptomania: The Egyptian Revival, a Recurring Theme in the History of Taste*. Manchester, UK: Manchester University Press, 1994.

Dacome, Lucia. *Malleable Anatomies: Models, Makers, and Material Culture in Eighteenth-Century Italy*. Oxford: Oxford University Press, 2017.

Dacre, Charlotte. *Zofloya, or, The Moor*. Oxford: Oxford University Press, 1997.

Davison, Carol Margaret, ed. *The Gothic and Death*. Manchester, UK: Manchester University Press, 2017.

Davys, Mary. *The Reform'd Coquet, Memoirs of Amoranda: Familiar Letters Betwixt a Gentleman and a Lady, and the Accomplish'd Rake.* Edited by Martha F. Bowden. Lexington: University Press of Kentucky, 1999.

Dawes, Ginny Redington, and Olivia Collings. *Georgian Jewellery 1714–1830.* Woodbridge, UK: ACC Art Books, 2019.

Dawson, Warren R., and P. H. K. Gray. *Catalogue of Egyptian Antiquities in the British Museum.* Vol. 1, *Mummies and Human Remains.* Department of Egyptian Antiquities, British Museum, 1968.

Day, W. G., and Kenneth Monkman. "The Skull." *The Shandean* 10 (1998): 45–79.

Daybell, James. *The Material Letter in Early Modern England: Manuscript Letters and the Culture and Practices of Letter-Writing, 1512–1635.* Early Modern Literature in History. New York: Palgrave Macmillan, 2012.

Defoe, Daniel. *The Life and Actions of Moll Flanders.* London: T. Read, 1723.

———. *Moll Flanders.* London: Penguin, 1989.

Degenaar, Marjolein. *Molyneux's Problem: Three Centuries of Discussion on the Perception of Forms.* New York: Springer, 1996.

DeLorme, Maureen. *Mourning Art & Jewelry.* Atglen, PA: Schiffer, 2004.

Dent, Edward. *Everlasting Blessedness: A Sermon Preached at the Funeral of . . . Mr. William Baker* London: Printed for W. Marshall, 1692.

A Descriptive Catalogue (Giving a Full Explanation) of Rackstrow's Museum [. . .]. ECCO Database. London, 1784.

A Descriptive Catalogue of the Sevveral Superb and Magnificent Pieces of Mechanism and Jewellery, Exhibited in the Museum, at Spring Gardens, Charing Cross. London, 1773.

Deutsch, Helen. "'Alas Poor Yorick!': Jonathan Swift, Madness and Fashionable Science." In *Disease and Death in Eighteenth-Century Literature and Culture: Fashioning the Unfashionable,* edited by Allan Ingram and Leigh Weatherall Dickson, 225–45. London: Palgrave Macmillian, 2016.

———. *Loving Dr. Johnson.* Chicago: University of Chicago Press, 2005.

Deutsch, Helen, and Mary Terrall, eds. *Vital Matters: Eighteenth-Century Views of Conception, Life, and Death.* Toronto: University of Toronto Press, 2012.

Dobson, Eleanor, and Nichola Tonks, eds. *Ancient Egypt in the Modern Imagination: Art, Literature and Culture.* London: Bloomsbury, 2020.

Dobson, Jessie. "Some Eighteenth Century Experiments in Embalming." *Journal of the History of Medicine and Allied Sciences* 8, no. 4 (1953): 431–41.

Dobson, Mary J. *Contours of Death and Disease in Early Modern England.* Cambridge Studies in Population, Economy, and Society in Past Time 29. New York: Cambridge University Press, 1997.

Dollimore, Jonathan. *Death, Desire, and Loss in Western Culture.* New York: Routledge, 1998.

Doody, Margaret Anne. *Frances Burney: The Life in the Works.* New Brunswick, NJ: Rutgers University Press, 1988.

———. *A Natural Passion: A Study of the Novels of Samuel Richardson.* Oxford: Clarendon Press, 1974.

———. *The True Story of the Novel.* New Brunswick, NJ: Rutgers University Press, 1996.

Doolittle, Thomas. *A Sermon on Eyeing of Eternity, so That It May Have Its Due Influence upon Us in All We Do.* 2nd ed. London, 1755.

Doran, John. *"Mann" and Manners at the Court of Florence, 1740–1786 Founded on the Letters of Horace Mann to Horace Walpole.* Vol. 2. London: Richard Bentley and Son, 1876.

Ducheyne, Steffen, ed. *Reassessing the Radical Enlightenment*. New York: Routledge, 2017.

Ebenstein, Joanna. *The Anatomical Venus: Wax, God, Death & the Ecstatic*. London: Thames & Hudson, 2016.

Elliott, Kamilla. *Portraiture and British Gothic Fiction: The Rise of Picture Identification, 1764–1835*. Baltimore: Johns Hopkins University Press, 2012.

Engel, Laura. "Epilogue: Body Double—Katharine Hepburn at Madame Tussauds." In *Intimacy and Celebrity in Eighteenth-Century Literary Culture: Public Interiors*, 293–98. Cham, Switzerland: Palgrave Macmillan, 2018.

Evans, J. "On the Utility of Paintings." *Scots Magazine*, July 1795.

Fennetaux, Ariane. "Fashioning Death/Gendering Sentiment: Women and the Material Culture of Mourning in Late Eighteenth-Century Europe." In *Women and the Material Culture of Death*, edited by Maureen Daly Goggin and Beth Fowkes Tobin. New York: Routledge, 2016.

Fennetaux, Ariane, Amélie Junqua, and Sophie Vasset, eds. *The Afterlife of Used Things: Recycling in the Long Eighteenth Century*. Routledge Studies in Cultural History 26. New York: Routledge, 2014.

Fielding, Henry. *Amelia*. London: Penguin, 1987.

Fielding, Sarah. *The Adventures of David Simple*. Lexington: University Press of Kentucky, 1998.

Finucane, R. C. "Sacred Corpse, Profane Carrion: Social Ideals and Death Rituals in the Later Middle Ages." In *Mirrors of Mortality: Studies in the Social History of Death*, edited by Joachim Whaley, 40–60. New York: St. Martin's Press, 1981.

Foucault, Michel. *The Birth of the Clinic: An Archaeology of Medical Perception*. New York: Vintage, 1994.

———. *The History of Sexuality*. Vol. 1, *An Introduction*. New York: Vintage, 1990.

Franceschina, John Charles. *Rape, Incest, Murder! The Marquis de Sade on Stage*. Vol. 3, *Asylum Plays*. Albany, GA: BearManor Media, 2015.

Frank, Robin Jaffee. *Love and Loss: American Portrait and Mourning Miniatures*. New Haven, CT: Yale University Art Gallery, 2000.

Fraser, Antonia. *Cromwell: The Lord Protector*. New York: Alfred A. Knopf, 1973.

Friedman, Alan Warren. *Fictional Death and the Modernist Enterprise*. Cambridge, UK: Cambridge University Press, 1995.

Friedman, Emily C. *Reading Smell in Eighteenth-Century Fiction*. Transits: Literature, Thought & Culture, 1650–1850. Lewisburg, PA: Bucknell University Press, 2016.

Fritz, Paul S. "From 'Public' to 'Private': The Royal Funerals in England, 1500–1830." In *Mirrors of Mortality*, edited by Joachim Whaley, 61–79. New York: Routledge, 1982.

———. "The Undertaking Trade in England: Its Origins and Early Development, 1660–1830." *Eighteenth-Century Studies* 28 (1994–1995): 241–53.

Fumerton, Patricia. *Cultural Aesthetics: Renaissance Literature and the Practice of Social Ornament*. Chicago: University of Chicago Press, 1991.

Gallagher, Catherine, and Thomas Walter Laqueur. *The Making of the Modern Body: Sexuality and Society in the Nineteenth Century*. Representations (Berkeley, Calif.) 14. Berkeley: University of California Press, 1987.

Gallagher, Shaun. *How the Body Shapes the Mind*. Oxford: Clarendon Press, 2005.

Gaukroger, Stephen. *The Natural and the Human: Science and the Shaping of Modernity, 1739–1841*. Oxford: Oxford University Press, 2016.

Gélis, Jacques. "The Saints of the Entrails and the Bowels of the Earth." In *Bellies, Bowels and Entrails in the Eighteenth Century*, 311–31. Manchester, UK: Manchester University Press, 2018.

Gevirtz, Karen Bloom. *Life After Death: Widows and the English Novel, Defoe to Austen*. Newark: University of Delaware Press, 2005.

Gibbon, Edward. *The History of the Decline and Fall of the Roman Empire*. Vol. 3. 6 vols. London, 1776.

Gibson, William. *The Church of England, 1688–1832: Unity and Accord*. London: Routledge, 2001.

Gittings, Clare. *Death, Burial, and the Individual in Early Modern England*. London: Croom Helm, 1984.

Goggin, Maureen Daly, and Beth Fowkes Tobin, eds. *Women and the Material Culture of Death*. Farnham, UK: Ashgate, 2013.

Gordon, Alexander. *An Essay Towards Explaining the Hieroglyphical Figures, on the Coffin of the Ancient Mummy Belonging to Capt. William Lethieullier*. London, 1737.

———. *Essay Towards Explaining the Hieroglyphical Figures, on the Egyptian Mummy, in the Museum of Doctor Mead, Physician in Ordinary to His Majesty*. London, 1737.

Gordon, Bruce, and Peter Marshall. *The Place of the Dead: Death and Remembrance in Late Medieval and Early Modern Europe*. Cambridge, UK:Cambridge University Press, 2000.

Goring, Paul. *The Rhetoric of Sensibility in Eighteenth-Century Culture*. Cambridge, UK: Cambridge University Press, 2005.

Gough, Richard. *Sepulchral Monuments in Great Britain: Applied to Illustrate the History of Families, Manners, Habits and Arts [. . .]*. 2 vols. T. Payne and Son, 1787.

Greenhill, Thomas. *Nekrokedeia: Or, the Art of Embalming: Wherein Is Shewn the Right of Burial [. . .]*. Eighteenth Century Collections Online. London: Printed for the author, 1705.

Gregory, Jeremy. "Religion: Faith in the Age of Reason." *Journal for Eighteenth-Century Studies* 34, no. 4 (2011): 435–43.

———. "Religion in the Age of Enlightenment: Putting John Wesley in Context." In *Religion in the Age of Enlightenment*. Vol. 2. New York: AMS Press, 2010.

Griffin, Elizabeth. *The Friends; or, The Contrast between Virtue and Vice: A Tale. Designed for the Improvement of Youth*. London: Printed for Crosby and Letterman, No. 4, Stationers Court, Paternoster Row, 1799.

Griffith, Elizabeth. *The Delicate Distress*. Lexington: University Press of Kentucky, 2015.

Grootenboer, Hanneke. *Treasuring the Gaze: Intimate Vision in Late Eighteenth-Century Eye Miniatures*. Chicago: University of Chicago Press, 2012.

Gross, Jonathan David *The Life of Anne Damer: Portrait of a Regency Artist*. Lanham, MD: Lexington Books, 2013.

Guerrini, Anita. "Anatomists and Entrepreneurs in Early Eighteenth-Century London." *Journal of the History of Medicine and Allied Sciences* 59 (2004): 219–39.

———. "The Value of a Dead Body." In *Vital Matters: Eighteenth Century Views of Conception, Life, and Death*. Toronto: University of Toronto Press, 2012.

Habenstein, Robert W., and William M. Lamers. *The History of American Funeral Directing*. Milwaukee: Bulfin, 1955.

Hadley, John. "An Account of a Mummy, Inspected at London 1763." *Philosophical Transactions (1683–1775)* 54 (1764): 1–14.

Hallam, E., and Jennifer Lorna Hockey. *Death, Memory and Material Culture*. Oxford: Berg, 2000.

Hancock, E. Geoffrey, Nick Pearce, and Mungo Campbell. *William Hunter's World: The Art and Science of Eighteenth-Century Collecting*. The Histories of Material Culture and Collecting, 1700–1950. London: Routledge, 2017.

Harvey, Anthony Ernest, and Richard Mortimer, eds. *The Funeral Effigies of Westminster Abbey*. Woodbridge, UK: Boydell, 2003.

Harvey, Elizabeth D. *Sensible Flesh: On Touch in Early Modern Culture*. Philadelphia: University of Pennsylvania Press, 2003.

Haviland, Thomas N., and Lawrence Charles Parish. "A Brief Account of the Use of Wax Models in the Study of Medicine." *Journal of the History of Medicine and Allied Sciences* 25, no. 1 (1970): 52–75.

Hayward, John. *The Horrors and Terrors of the Hour of Death: And Day of Judgment . . . With Holy Directions to Die Well. And Also the Great Danger of a Death-Bed Repentance. . . .* Eighteenth Century Collections Online. London: Printed for Robert Gifford, 1707.

Haywood, Eliza. *The History of Miss Betsy Thoughtless*. Oxford: Oxford University Press, 1997.

Hempton, David. *The Church in the Long Eighteenth Century*. Vol. 5. The I. B. Tauris History of the Christian Church. London: I. B. Tauris, 2011.

Hertz, Niel. "Dr. Johnson's Forgetfulness, Descartes' Piece of Wax." *Eighteenth-Century Life* 16 (1992): 167–81.

Hervey, James. *Meditations Among the Tombs: In a Letter to a Lady*. 12th ed. Vol. 2. Dublin, 1758.

———. *Meditations and Contemplations*. London: J. F. and C. Rivington, 1779.

Hill, Robert. *The Pathway to Prayer and Pietie Contayning an Exposition of the Lord's Prayer*. London, 1616.

Hoeveler, Diane Long. *Gothic Feminism: The Professionalization of Gender from Charlotte Smith to the Brontës*. University Park: Pennsylvania State University Press, 1998.

Holm, Christiane. "Sentimental Cuts: Eighteenth-Century Mourning Jewelry with Hair." *Eighteenth-Century Studies* 38, no. 1 (2004): 139–43.

Hooker, Richard. *Of the Lawes of Ecclesiastical Politie*. 2 vols. London: Printed and sold by Benjamin Bragg, at the Blue Ball in Avemary-Lane, 1705.

Houlbrooke, Ralph A. *Death, Religion, and the Family in England, 1480–1750*. Oxford Studies in Social History. Oxford: Clarendon Press, 1998.

———. *Death, Ritual, and Bereavement*. London: Routledge, 1989.

Hume, David. *A Treatise of Human Nature*. Project Gutenberg, 1738. https://www.gutenberg.org/files/4705/4705-h/4705-h.htm.

Hunter, John. *Essays and Observations on Natural History, Anatomy, Physiology, Psychology, and Geology*. Vol. 1. London: J. Van Voorst, 1861.

Hunter, Margaret. "Mourning Jewellery: A Collector's Account." *Costume* 27, no. 1 (1993): 9–26.

Hunter, William. *Hunter's Lectures of Anatomy*. Edited by Nell Dowd. Amsterdam: Elsevier,1972.

Hussey, Kristin D. "Seen and Unseen: The Representation of Visible and Hidden Disease in the Waxworks of Joseph Towne at the Gordon Museum." *19: Interdisciplinary Studies in the Long Nineteenth Century* 24 (2017).

Inchbald, Elizabeth. *A Simple Story*. Oxford: Oxford University Press, 1998.

Ingram, Allan, and Leigh Wetherall Dickson. *Disease and Death in Eighteenth-Century Literature and Culture: Fashioning the Unfashionable*. Palgrave Studies in Literature, Science and Medicine. London: Palgrave Macmillan, 2016.

Ingrassia, Catherine. *Authorship, Commerce, and Gender in Early Eighteenth-Century England: A Culture of Paper Credit*. New York: Cambridge University Press, 1998.

Isenberg, Nancy, and Andrew Burstein, eds. *Mortal Remains: Death in Early America*. Philadelphia: University of Pennsylvania Press, 2002.

Israel, Jonathan. *Radical Enlightenment: Philosophy and the Making of Modernity, 1650–1750*. Oxford: Oxford University Press, 2001.

Jacob, Margaret C. *The Radical Enlightenment—Pantheists, Freemasons, and Republicans*. London: George Allen and Unwin, 1981.

———. *The Secular Enlightenment*. Princeton, NJ: Princeton University Press, 2019.
Johnson, Mark. *The Meaning of the Body: Aesthetics of Human Understanding*. Chicago: University of Chicago Press, 2007.
Jones, David J. *Gothic Effigy: A Guide to Dark Visibilities*. Manchester, UK: Manchester University Press, 2018.
———. *Gothic Machine: Textualities, Pre-Cinematic Media and Film in Popular Visual Culture, 1670–1910*. Gothic Literary Studies. Cardiff: University of Wales Press, 2011.
Jones, Emrys, and Victoria Joule, eds. *Intimacy and Celebrity in Eighteenth-Century Literary Culture: Public Interiors*. Cham, Switzerland: Palgrave Macmillan, 2018.
Jordanova, Ludmilla. *Sexual Visions: Images of Gender in Science and Medicine Between the Eighteenth and Twentieth Centuries*. Science and Literature. Madison: University of Wisconsin Press, 1993.
Jupp, Peter C. *From Dust to Ashes: Cremation and the British Way of Death*. Basingstoke, UK: Palgrave Macmillan, 2006.
Jupp, Peter C., and Clare Gittings. *Death in England: An Illustrated History*. New Brunswick, NJ: Rutgers University Press, 2000.
Jupp, Peter C., and Glennys Howarth, eds. *The Changing Face of Death: Historical Accounts of Death and Disposal*. Basingstoke: Macmillan, 1997.
Justice, Elizabeth. *Amelia, or, the Distress'd Wife: A History Founded on Real Circumstances. By a Private Gentlewoman*. ECCO. London, 1751.
Kareem, Sarah Tindal. *Eighteenth-Century Fiction and the Reinvention of Wonder*. Oxford: Oxford University Press, 2014.
Kelly, Alison. "The Storied Urn." In *Decorative Wedgwood*. London: Country Life, 1965.
Kemp, Martin, and Marina Wallace. *Spectacular Bodies: The Art and Science of the Human Body from Leonardo to Now*. London: Hayward Gallery, 2000.
Kettlewell, John. *Death Made Comfortable: Or the Way to Die Well, Consisting of Directions for an Holy and an Happy Death*. Eighteenth Century Collections Online. London: Printed for A. and J. Churchill, 1702.
Kibbie, Anne Louise. "Realism and Decay in Wax." *Configurations* 25, no. 2 (2017): 165–87.
Knight, Roger. *The Pursuit of Victory: The Life and Achievement of Horatio Nelson*. New York: Basic Books, 2005.
Kobler, John. *The Reluctant Surgeon: A Biography of John Hunter*. New York: Doubleday, 1960.
Kooijmans, Luuc. *Death Defied: The Anatomy Lessons of Frederik Ruysch*. Translated by Diane Webb. Boston: Brill, 2011.
Korkow, Cory. *British Portrait Miniatures: The Cleveland Museum of Art*. London: D. Giles, 2013.
La Belle Assemblée, or, Bell's Court and Fashionable Magazine, August 1, 1817.
Ladies Magazine, Or, The Universal Entertainer, August 22, 1752.
The Lady's Delight, or Accomplish'd Female Instructor: Being a Very Useful Companion for Ladies, Gentlewomen, and Others [. . .]. Eighteenth Century Collections Online. London: Printed for James Hodges, 1730.
Lake, Crystal B. *Artifacts: How We Think and Write About Found Objects*. Baltimore, MD: Johns Hopkins University Press, 2020.
Lakoff, George, and Mark Johnson. *Philosophy in the Flesh: The Embodied Mind and Its Challenge to Western Thought*. New York: Basic Books, 1999.
Laqueur, Thomas W. *The Work of the Dead: A Cultural History of Mortal Remains*. Princeton, NJ: Princeton University Press, 2015.

Larson, Frances. *Severed: A History of Heads Lost and Heads Found*. New York: Liveright, 2014.

Latham, Robert, and William Matthews, eds. *The Diary of Samuel Pepys*. 11 vols. Oakland: University of California Press, 1976.

Latour, Bruno. *Reassembling the Social: An Introduction to Actor–Network–Theory*. Oxford: Oxford University Press, 2005.

———. *We Have Never Been Modern*. Translated by Catherine Porter. Cambridge, MA: Harvard University Press, 1993.

Lawlor, Clark. *Consumption and Literature: The Making of the Romantic Disease*. Basingstoke, UK: Palgrave Macmillan, 2006.

Lehner, Ulrich L. *The Catholic Enlightenment: The Forgotten History of a Global Movement*. Oxford: Oxford University Press, 2016.

Lewis, Jayne. "'A Lock of Thy Bright Hair': The Enlightenment's Milton and Our Auratic Material." *Humanities* 4, no. 4 (2015): 797–817.

Lightfoot, D. Tulla. *The Culture and Art of Death in 19th-Century America*. Jefferson, NC: McFarland & Company, 2019.

Lindsay, Suzanne G. *Funerary Arts and Tomb Cult: Living with the Dead in France, 1750–1870*. Farnham, UK: Ashgate, 2012.

Lindsay, Suzanne Glover. "Mummies and Tombs: Turenne, Napoleon, and Death Ritual." *The Art Bulletin* 82, no. 3 (2000): 476–502.

Lipski, Jakub. *Painting the Novel: Pictorial Discourse in Eighteenth-Century English Fiction*. New York: Routledge, 2017.

Litten, Julian. *The English Way of Death: The Common Funeral Since 1450*. London: Robert Hale, 1991.

———. "Tombs Fit for Kings: Some Burial Vaults of the English Aristocracy and Landed Gentry of the Period 1650–1850." *Church Monuments* 14 (1999): 104–28.

Llewellyn, Nigel. *The Art of Death: Visual Culture in the English Death Ritual, c. 1500–c. 1800*. London: Published in association with the Victoria and Albert Museum by Reaktion Books, 1991.

Lloyd, Henry Martyn, ed. *The Discourse of Sensibility: The Knowing Body in the Enlightenment*. New York: Springer, 2013.

Lloyd, Stephen. *Portrait Miniatures from the National Galleries of Scotland*. Edinburgh: Scottish National Portrait Gallery, 2004.

Locke, John. *Locke's Essays: An Essay Concerning Human Understanding, and A Treatise on the Conduct of the Understanding*. Philadelphia: James Kay, Jr. & Brother, 1844. doi:10.1037/11955-002.

The London and Dublin Magazine: Or, Gentleman's Monthly Intelligencer. Dublin: Printed by S. Powell, 1734.

Lovejoy, Bess. "How Did Bits of Percy Shelley's Skull End Up in the New York Public Library?" In *Atlas Obscura*, July 8, 2013. https://www.atlasobscura.com/articles/shelley-skull-fragments-at-nypl.

———. *Rest in Pieces: The Curious Fates of Famous Corpses*. New York: Simon & Schuster, 2013.

Luthi, Anne Louise. *Sentimental Jewellery*. Buckinghamshire, UK: Shire Publications, 2007.

Lutz, Deborah. *The Brontë Cabinet: Three Lives in Nine Objects*. New York: W. W. Norton, 2016.

———. *Relics of Death in Victorian Literature and Culture*. Cambridge Studies in Nineteenth-Century Literature and Culture. Cambridge, UK: Cambridge University Press, 2015.

Mack, Phillis. "The Senses in Religion: Listening to God in the Eighteenth Century." In *A Cultural History of the Senses in the Age of Enlightenment*, edited by Anne Vila. London: Bloomsbury, 2014.

Mackay, John Macdonald. *A Miscellany Presented to John Macdonald Mackay, LL. D., July, 1914*. Edited by Oliver Elton. Liverpool: University Press, 1914.

Maerker, Anna Katharina. *Model Experts: Wax Anatomies and Enlightenment in Florence and Vienna, 1775–1815*. Manchester, UK: Manchester University Press, 2011.

Marsden, Jean. "Tragedy and Varieties of Serious Drama." In *A Companion to Restoration Drama*, edited by Susan J. Owen. Oxford: Blackwell Publishers, 2002.

Marshall, Peter. *Beliefs and the Dead in Reformation England*. Oxford: Oxford University Press, 2002.

Marshall, Peter, and Bruce Gordon, eds. *The Place of the Dead: Death and Remembrance in Late Medieval and Early Modern Europe*. Cambridge, UK: Cambridge University Press, 2000.

Matthews, Samantha. *Poetical Remains: Poets' Graves, Bodies, and Books in the Nineteenth Century*. OUP E-Books. Oxford: Oxford University Press, 2004.

McAllister, David. *Imagining the Dead in British Literature and Culture, 1790–1848*. London: Palgrave Macmillan, 2018.

McFarland, Thomas. *The Masks of Keats: The Endeavour of a Poet*. Oxford: Oxford University Press, 2000.

McGuire, Kelly. *Dying to Be English: Suicide Narratives and National Identity, 1721–1814*. New York: Routledge, 2012.

McInelly, Brett C., ed. *Religion in the Age of Enlightenment*. Vol. 2. New York: AMS Press, 2010.

McKeon, Michael. *The Origins of the English Novel, 1600–1740*. Baltimore: Johns Hopkins University Press, 1987.

McMains, H. F. *The Death of Oliver Cromwell*. Lexington: University Press of Kentucky, 2000.

McManners, John. *Death and the Enlightenment: Changing Attitudes to Death Among Christians and Unbelievers in Eighteenth-Century France*. Oxford: Clarendon Press, 1981.

McMaster, Juliet. *Reading the Body in the Eighteenth-Century Novel*. New York: Palgrave Macmillan, 2004.

McWhorter, Ellen. "'Deeply and Suddenly Touched': Homoerotics in James Joyce's Exiles." *James Joyce Quarterly* 53, no. 1–2 (2015): 23–36.

Meades, Anna. *The History of Sir William Harrington. Written Some Years Since, And Revised and Corrected by the Late Mr. Richardson [. . .]*,. 4 vols. London, 1772.

Messbarger, Rebecca Marie. *The Lady Anatomist: The Life and Work of Anna Morandi Manzolini*. Chicago: University of Chicago Press, 2010.

Molesworth, Jesse. *Chance and the Eighteenth-Century Novel: Realism, Probability, Magic*. Cambridge, UK: Cambridge University Press, 2014.

Montagu, Mary Wortley. *Mary Wortley Montagu, Letters Written during Her Travels in Europe, Asia and Africa*. Sarah Goddard, printer, 1766.

Moore, Wendy. *The Knife Man: Blood, Body Snatching, and the Birth of Modern Surgery*. New York: Broadway Books, 2005.

Morgan, Michael J. *Molyneux's Question: Vision, Touch, and the Philosophy of Perception*. Cambridge, UK: Cambridge University Press, 1977.

Motion, Andrew. *Keats*. New York: Farrar, Straus and Giroux, 1998.

Mullan, John. *Sentiment and Sociability: The Language of Feeling in the Eighteenth Century*. Oxford: Clarendon Press, 1988.

Napier, Elizabeth R. *Falling into Matter: Problems of Embodiment in English Fiction from Defoe to Shelley*. Toronto: University of Toronto Press, 2012.

Nehama, Sarah, and Anne E. Bentley. *In Death Lamented: The Tradition of Anglo-American Mourning Jewelry*. Boston: Massachusetts Historical Society, 2012.

Neve, Philip. *A Narrative of the Disinterment of Milton's Coffin [. . .]*. Eighteenth Century Collections Online. London, 1790.

Newby, Zahra, and Ruth E. Toulson, eds. *The Materiality of Mourning: Cross-Disciplinary Perspectives*. New York: Routledge, 2018.

Newton, Isaac. *The Mathematical Principles of Natural Philosophy and His System of the World (Principia)*. Translated by Andrew Motte. Berkeley: University of California Press, 1966.

Nokes, David. *Samuel Johnson: A Life*. New York: Henry Holt, 2009.

Oliver, Kathleen M. *Narrative Mourning: Death and Its Relics in the Eighteenth-Century British Novel*. Lewisburg, PA: Bucknell University Press, 2020.

———. "'With My Hair in Crystal': Mourning Clarissa." *Eighteenth-Century Fiction* 23, no. 1 (2010): 35–60.

Oram, Richard. "The Harry Ransom Center Magazine." *The Locks of Ages: The Leigh Hunt Hair Collection* (blog), January 13, 2001. https://sites.utexas.edu/ransomcentermagazine/2011/01/13/locks-of-ages-the-leigh-hunt-hair-collection.

Owen, Susan J. *A Companion to Restoration Drama*. Oxford: Blackwell, 2002.

Packham, Catherine. *Eighteenth-Century Vitalism: Bodies, Culture, Politics*. London: Palgrave Macmillan, 2012.

Palmer, Charlotte. *It Is, and It Is Not a Novel*. Eighteenth Century Collections Online. London, 1792.

Panzanelli, Roberta, and Julius von Schlosser. *Ephemeral Bodies: Wax Sculpture and the Human Figure*. Los Angeles: Getty Research Institute, 2008.

Panzanelli, Roberta., Eike D. Schmidt, and Kenneth D. S. Lapatin. *The Color of Life: Polychromy in Sculpture from Antiquity to the Present*. Los Angeles: J. Paul Getty Museum, 2008.

Parisot, Eric. *Graveyard Poetry: Religion, Aesthetics and the Mid-Eighteenth-Century Poetic Condition*. British Literature in Context in the Long Eighteenth Century. Farnham, UK: Ashgate, 2013.

———. "The Work of Feeling in James Hervey's Meditations Among the Tombs (1746)." *Parergon* 31, no. 2 (2014): 121–35.

Park, Julie. "Pains and Pleasures of the Automaton: Frances Burney's Mechanics of Coming Out." *Eighteenth-Century Studies* 40, no. 1 (2006): 23–49.

———. *The Self and It: Novel Objects in Eighteenth-Century England*. Stanford, CA: Stanford University Press, 2010.

Parsons, Brian. *The Evolution of the British Funeral Industry in the 20th Century: From Undertaker to Funeral Director*. Emerald Studies in Death and Culture. Bingley, UK: Emerald, 2018.

Paterson, Mark. *Seeing with the Hands: Blindness, Vision, and Touch After Descartes*. Edinburgh: Edinburgh University Press, 2016.

———. *The Senses of Touch: Haptics, Affects, and Technologies*. Oxford: Berg, 2007.

Pearson, George. *Observations and Experiments for Investigating the Chymical History of the Tepid Springs of Buxton [. . .]*. London: Printed for J. Johnson, 1784.

Pearson, John. *An Exposition of the Creed*. 12th ed. London: Printed by Roger Daniel for John Williams [. . .], 1741.

Pembroke, Anne Clifford Herbert, and D. J. H. Clifford. *The Diaries of Lady Anne Clifford*. Stroud: Sutton, 1990.

Phillippy, Patricia Berrahou. *Women, Death, and Literature in Post-Reformation England*. Cambridge, UK: Cambridge University Press, 2002.

Pilbeam, Pamela M. *Madame Tussaud and the History of Waxworks*. London: Hambledon Press, 2003.

Plumly, Stanley. *Posthumous Keats: A Personal Biography*. New York: W. W. Norton., 2008.

Pointon, Marcia. *Brilliant Effects: A Cultural History of Gem Stones and Jewellery*. New Haven, CT: Yale University Press, 2010.

———. "Materializing Mourning: Hair, Jewellery and the Body." In *Classic and Modern Writings on Fashion*. Oxford ,UK: Berg, 2009.

Pointon, Marcia R. *Hanging the Head: Portraiture and Social Formation in Eighteenth-Century England*. New Haven, CT: Published for the Paul Mellon Centre for Studies in British Art by Yale University Press, 1993.

Polson, C. J., R. P. Brittain, and T. K. Marshall. *The Disposal of the Dead*. New York: Philosophical Library, 1953.

Ponsonby, Henrietta Frances Spencer, Countess of Bessborough. *Lady Bessborough and Her Family Circle*. Edited by Vere Brabazon Ponsonby, 9th earl of Bessborough, and Arthur Aspinall. London: J. Murray, 1940.

Pope, Rodney J. "The Eighteenth Century Church in the Wirral." MA thesis, University of Wales, Lampeter, 1971

Porter, Roy. *English Society in the Eighteenth Century*. London: Penguin, 1982.

———. *Enlightenment: Britain and the Creation of the Modern World*. New York: Penguin, 2001.

———. *Flesh in the Age of Reason*. New York: W. W. Norton, 2004.

Porter, Stephen. "Death and Burial in a London Parish: St Mary Woolnoth 1653–99." *London Journal* 8, no. 1 (1982): 76–80.

———. "Death and Burial in Seventeenth-Century Oxford." *Oxfordshire Local History* 1 (1980): 2–7.

Powell, Margaret K., and Joseph. Roach. *A Cultural History of Hair in the Age of Enlightenment*. London: Bloomsbury, 2020.

Pringle, John. *Observations on the Diseases of the Army*. The 6th ed., Corrected. ECCO. London: Printed for A. Millar and T. Cadell, D. Wilson, 1768.

———. "Some Experiments on Substances Resisting Putrefaction." *Philisophical Transactions* 47 (1749): 480–88.

Puckle, Bertram S. *Funeral Customs: Their Origin and Development*. London: Laurie, 1926.

Punday, Daniel. *Narrative Bodies: Toward a Corporeal Narratology*. New York: Palgrave Macmillan, 2003.

Purcell, Rosamond Wolff, and Stephen Jay Gould. *Finders, Keepers: Eight Collectors*. New York: W. W. Norton, 1992.

Purnell, Carolyn. *The Sensational Past: How the Enlightenment Changed the Way We Use Our Senses*. New York: W. W. Norton, 2017.

Pyke, E. J. *A Biographical Dictionary of Wax Modellers*. Oxford: Clarendon Press, 1973.

Rabb, Melinda Alliker. *Miniature and the English Imagination: Literature, Cognition, and Small-Scale Culture 1650–1765*. Cambridge, UK: Cambridge University Press, 2019.

Radcliffe, Ann. *The Mysteries of Udolpho*. Oxford: Oxford University Press, 1998.

———. *The Romance of the Forest*. Oxford: Oxford University Press, 2009.

———. *A Sicilian Romance*. Oxford: Oxford University Press, 1998.

Reeve, Clara. *The Old English Baron*. Oxford: Oxford University Press, 2008.

Reilly, David Robin. *Portrait Waxes: An Introduction for Collectors*. London: Batsford, 1953.

Reinarz, Jonathan, and Leonard Schwarz. "The Senses and the Enlightenment: An Introduction." *Journal for Eighteenth-Century Studies* 35, no. 4 (2012): 465–68.

Reynolds, Graham. *English Portrait Miniatures*. Library of English Art. Cambridge, UK: Cambridge University Press, 1988.

Richardson, Ruth. *Death, Dissection and the Destitute*. Chicago: University of Chicago Press, 2000.

Richardson, Ruth, and Brian Hurwitz. "Jeremy Bentham's Self Image: An Exemplary Bequest for Dissection." *British Medical Journal* 295, no. 6591 (1987): 195–98.

Richardson, Samuel. *Clarissa, or The History of a Young Lady*. 3rd ed. London, 1751.

———. *Early Works: 'Aesop's Fables', 'Letters Written to and for Particular Friends' and Other Works*. Edited by Alexander Pettit. Cambridge, UK: Cambridge University Press, 2012.

Richter, David H. *Ideology and Form in Eighteenth-Century Literature*. Lubbock: Texas Tech University Press, 1999.

Riskin, Jessica. *Science in the Age of Sensibility: The Sentimental Empiricists of the French Enlightenment*. Chicago: University of Chicago Press, 2002.

Roberts, Andrew. *Napoleon: A Life*. London: Penguin, 2015.

Robinson, Mary. *The Widow, Or a Picture of Modern Times: A Novel, in a Series of Letters, in Two Volumes*. London, 1794.

Roe, Nicholas. *John Keats: A New Life*. New Haven, CT: Yale University Press, 2012.

Ross, Ian Campbell. *Laurence Sterne: A Life*. Oxford: Oxford University Press, 2001.

Rousseau, Jean-Jacques. *Émile*. London: J. M. Dent, 1966.

Royal Archaeological Institute of Great Britain and Ireland. *The Embalmed Head of Oliver Cromwell: In the Possession of the Rev. H.R. Wilkinson: Exhibited Before the Royal Archaeological Institute [. . .]*. London: The Institute, 1911.

Royal Commission on Historical Manuscripts, and John Perceval Earl of Egmont. *Manuscripts of the Earl of Egmont: Diary of Viscount Percival, Afterwards First Earl of Egmont*. Historical Manuscripts Commission 63. London: HMSO, 1920.

Runte, Roseann. "Dying Words: The Vocabulary of Death in Three Eighteenth-Century English and French Novels." *Canadian Review of Comparative Literature* 6, no. 4 (1979): 360–68.

Sabor, Peter, and Betty A. Schellenberg, eds. *Samuel Richardson in Context*. Literature in Context. Cambridge, UK: Cambridge University Press, 2019.

Salmon, William. *Synopsis Medicinæ, or, A Compendium of Astrological, Galenical, & Chymical Physick [. . .]*. 2nd ed. London: Printed by W. Godbid, 1681.

Sandberg, Mark. *Living Pictures, Missing Persons: Mannequins, Museums, and Modernity*. Princeton, NJ: Princeton University Press, 2002.

Sandy, Mark. *Romanticism, Memory, and Mourning*. New York: Routledge, 2013.

Scarisbrick, Diana. *Rings: Jewelry of Power, Love and Loyalty*. London: Thames & Hudson, 2014.

———. *Rings: Symbols of Wealth, Power, and Affection*. New York: Abrams, 1993.

Scarry, Elaine. "Donne: 'But yet the body is his booke.'" In *Literature and the Body: Essays on Populations and Persons*, 70–105. Baltimore, MD: Johns Hopkins University Press, 1988.

———. *Literature and the Body: Essays on Populations and Persons*. Selected Papers from the English Institute New Series 12. Baltimore: Johns Hopkins University Press, 1990.

Schaeffer, Neil. *The Marquis de Sade: A Life*. Cambridge, MA: Harvard University Press, 1999.

Schor, Esther. *Bearing the Dead: The British Culture of Mourning from the Enlightenment to Victoria*. Princeton, NJ: Princeton University Press, 1999.

Schulman, Alex. *The Secular Contract: The Politics of Enlightenment*. London: Bloomsbury, 2013.

Shapira, Yael. *Inventing the Gothic Corpse: The Thrill of Human Remains in the Eighteenth-Century Novel*. London: Palgrave Macmillan, 2018.

Shaw, Jane. *Miracles in Enlightenment England*. New Haven, CT: Yale University Press, 2006.

Shelley, Mary. *Frankenstein*. New York: W. W. Norton, 2011.

Sheridan, Frances Chamberlaine, Patricia Köster, and Jean Coates Cleary. *Memoirs of Miss Sidney Bidulph*. Oxford World's Classics. Oxford: Oxford University Press, 1995.

Sheumaker, Helen. *Love Entwined: The Curious History of Hairwork in America*. Philadelphia: University of Pennsylvania Press, 2007.

Silver, Sean. *The Mind Is a Collection: Case Studies in Eighteenth-Century Thought*. Philadelphia: University of Pennsylvania Press, 2015.

Smith, Adam. "Chapter I: Of Sympathy, Section I: Of the Sense of Propriety." In *Theory of Moral Sentiments*, 1759.

Smith, Andrew. *Gothic Death 1740–1914: A Literary History*. Manchester Gothic. Manchester, UK: Manchester University Press, 2016.

Smith, Charlotte. *Celestina*. Edited by Loraine Fletcher. Broadview Editions. Toronto: Broadview Press, 2004.

———. *Desmond*. Edited by Janet M. Todd and Antje Blank. Peterborough, ON: Broadview Press, 2009.

———. *Desmond: A Novel, in Three Volumes*. Eighteenth Century Collections Online. London, 1792.

———. *Emmeline, The Orphan of the Castle*. Ontario: Broadview Press, 2003.

———. *Ethelinde, or The Recluse of the Lake*. Richmond, VA: Valancourt Books, 2017.

Smith, Mark. *Sensory History*. Oxford: Berg, 2007.

Smollett, Tobias. *The Adventures of Peregrine Pickle*. Oxford: Oxford University Press, 1969.

Sorkin, David. *The Religious Enlightenment: Protestants, Jews, and Catholics from London to Vienna*. Princeton, NJ: Princeton University Press, 2008.

Stabile, Susan M. *Memory's Daughters: The Material Culture of Remembrance in Eighteenth-Century America*. Ithaca, NY: Cornell University Press, 2004.

Stafford, Barbara Maria. *Body Criticism: Imaging the Unseen in Enlightenment Art and Medicine*. Cambridge, MA: MIT Press, 1991.

Stannard, David E. *The Puritan Way of Death: A Study in Religion, Culture, and Social Change*. New York: Oxford University Press, 1977.

Stefanacci, Davide. *The Vision of Death: Wax Sculpture in the Seventeenth Century*. North Charleston, SC: CreateSpace, 2018.

Stephens, Elizabeth. *Anatomy as Spectacle: Public Exhibitions of the Body from 1700 to the Present*. Representations: Health, Disability, Culture. Liverpool, UK: Liverpool University Press, 2011.

Sterckx, Marjan. "Pride and Prejudice: Eighteenth Century Women Sculptors and Their Material Practices." In *Women and the Material Culture of Death*, edited by Maureen Daly Goggin and Beth Fowkes Tobin, 283–310. Farnham, UK: Ashgate, 2013.

Sterne, Laurence. *The Life and Opinions of Tristram Shandy, Gentleman*. Edited by Melvyn New. London: Penguin, 2003.

Stewart, Carol. *The Eighteenth-Century Novel and the Secularization of Ethics*. New York: Routledge, 2010.

Stewart, Garrett. *Death Sentences: Styles of Dying in British Fiction*. Cambridge, MA: Harvard University Press, 1984.

Stewart, Susan. *On Longing: Narratives of the Miniature, the Gigantic, the Souvenir, the Collection*. Durham, NC: Duke University Press, 1993.

Sugg, Richard. *Mummies, Cannibals, and Vampires: The History of Corpse Medicine from the Renaissance to the Victorians*. London: Routledge, 2016.

Swift, Jonathan. *"A Tale of a Tub" and Other Works*. Edited by Angus Ross and David Woolley. Oxford World's Classics. Oxford: Oxford University Press, 1999.

Syson, Luke, et. al. "Polychrome and Its Discontents: A History." In *Like Life: Sculpture, Color, and the Body*, edited by Kamilah Foreman and Elizabeth Franzen, 14–42. New York: Metropolitan Museum of Art, 2018.

Taylor, Derek. *Reason and Religion in Clarissa: Samuel Richardson and "the Famous Mr. Norris, of Bemerton."* London: Routledge, 2018.

Taylor, Jeremy. *The Rule and Exercises of Holy Dying [. . .]*. 24th ed. London, 1727.

Taylor, Lou. *Mourning Dress: A Costume and Social History*. Routledge Revivals. New York: Routledge, 1983.

Teacher, John H. *Catalogue of the Anatomical and Pathological Preparations of Dr. William Hunter*. Glasgow: James MacLehose and Sons, 1900.

Thompson, Helen. *Fictional Matter: Empiricism, Corpuscles, and the Novel*. Philadelphia: University of Pennsylvania Press, 2017.

Tingle, Elizabeth C., and Jonathan P. Willis, eds. *Dying, Death, Burial and Commemoration in Reformation Europe*. New York: Routledge, 2016.

Todd, Janet. *Gender, Art, and Death*. New York: Continuum, 1993.

Toulmin, Stephen. *Cosmopolis: The Hidden Agenda of Modernity*. Chicago: University of Chicago Press, 1990.

Trenchard, John. *A Collection of Scarce and Valuable Tracts, on the Most [. . .]*. Vol. 3. London, 1948.

Tunstall, Kate E. *Blindness and Enlightenment. An Essay. With a New Translation of Diderot's Letter on the Blind*. London: Continuum, 2011.

Turner, Daniel. *Preparation for Death Explained and Enforced. A Sermon, Occasioned by the Death of Mrs. Elizabeth Tomkins . . . Preached at the Time of Her Interment*. Abingdon: Printed by J. Stacey, 1769.

Van Leeuwen, Evert Jan. "Funeral Sermons and Graveyard Poetry: The Ecstasy of Death and Bodily Resurrection." *Journal for Eighteenth-Century Studies* 32, no. 3 (2009): 353–71.

Van Sant, Ann Jessie. *Eighteenth-Century Sensibility and the Novel: The Senses in Social Context*. Cambridge, UK: Cambridge University Press, 2004.

Verney, Frances Parthenope, and Margaret Maria Williams-Hay, Lady Verney. *Memoirs of the Verney Family*. Vol. 4. London: Longmans, Green, 1892.

Vila, Anne C. *Enlightenment and Pathology: Sensibility in the Literature and Medicine of Eighteenth-Century France*. Baltimore: Johns Hopkins University Press, 1998.

———. "Introduction: Powers, Pleasures, and the Perils of the Senses in the Enlightenment Era." In *A Cultural History of the Senses in the Age of Enlightenment*, edited by Anne C. Vila. London: Bloomsbury, 2014.

Voltaire. *Philosophical Letters, or Letter Regarding the English Nation*. Edited by John Leigh. Translated by Prudence Steiner. Indianapolis: Hackett, 2007.

Wahrman, Dror. *Making of the Modern Self: Identity and Culture in Eighteenth-Century England*. New Haven, CT: Yale University Press, 2004.

Walker, D. P. *The Decline of Hell: Seventeenth-Century Discussions of Eternal Torment*. Chicago: University of Chicago Press, 1964.

Walker, John Frederick. *Ivory's Ghosts: The White Gold of History and the Fate of Elephants*. New York: Grove Atlantic, 2010.

Wallace, Elizabeth Kowaleski. "Recycling the Sacred: The Wax Votive Object and the Eighteenth-Century Wax Baby Doll." In *The Afterlife of Used Things: Recycling in the Long Eighteenth Century*. New York: Routledge, 2014.

Walmsley, Peter. "'Live to Die, Die to Live': An Introduction." *Eighteenth Century Fiction* 21, no. 1 (2008): 1–11.

Walpole, Horace. *The Yale Edition of Horace Walpole's Correspondence*. Edited by Wilmarth S. Lewis. 48 vols. New Haven, CT: Yale University Press, 1937–1983.

Walpole, Horace, and Edwine M. Martz. *Horace Walpole's Correspondence: Additions and Corrections, Subject Index to Illustrations, Index of Horace Walpole's Correspondents, Chronological List of Letters*. Vol. 43 of *The Yale Edition of Horace Walpole's Correspondence*. New Haven, CT: Yale University Press, 1983.

Walsham, Alexandra. "Skeletons in the Cupboard: Relics after the English Reformation." *Past and Present* 206, no. S5 (2010): 121–43.

Wardrop, James, ed. *The Works of Matthew Baillie, M.D. To Which Is Prefixed an Account of His Life*. Vol. 1. London: Longman, Hurst, Rees, et. al., 1825.

Warner, Marina. *Phantasmagoria: Spirit Visions, Metaphors, and Media into the Twenty-First Century*. Oxford: Oxford University Press, 2008.

———. "Waxworks and Wonderlands." In *Visual Display: Culture Beyond Appearances*, edited by Lynne Cooke and Peter Wollen, 179–201. New York: New Press, 1995, 179–201.

Watt, Ian P. *The Rise of the Novel: Studies in Defoe, Richardson, and Fielding*. Berkeley: University of California Press, 1957.

"Wax Works at Westminster Abbey." *Scientific American* 80, no. 10 (1899): 148.

Wendt, Allan. "Clarissa's Coffin." *Philological Quarterly*, no. 39 (January 1960): 481–95.

Wennerstrom, Courtney. "Cosmopolitan Bodies and Dissected Sexualities: Anatomical Mis-Stories in Ann Radcliffe's Mysteries of Udolpho." *European Romantic Review* 16, no. 2 (2005): 193–207.

Westover, Paul. *Necromanticism: Traveling to Meet the Dead 1750–1860*. London: Palgrave Macmillan, 2014.

Wetmore, Alex. *Men of Feeling in Eighteenth-Century Literature: Touching Fiction*. London: Palgrave Macmillan, 2013.

Whaley, Joachim. *Mirrors of Mortality: Studies in the Social History of Death*. Routledge Revivals. Abingdon: Routledge, 2011.

Whiting, Patricia A. "Women's Magazines and the Representation of Death in Eighteenth-Century England." PhD diss., University of Ottawa, 1999.

Wilkins, William. *A Specimen of Embalming the Dead Intire without Disbowelling*, 1690.

Williams, Helen. "'Alas, Poor YORICK!': Sterne's Iconography of Mourning." *Eighteenth Century Fiction* 28, no. 2 (2015).

Wollstonecraft, Mary. *Letters Written During a Short Residence in Sweden, Norway, and Denmark*. Edited by Tone Brekke and Jon Mee. Oxford World's Classics. Oxford: Oxford University Press, 2009.

Wright, George. *Pleasing Melancholy: Or a Walk Among the Tombs in a Country Church Yard, in the Stile & Manner of Hervey's Meditations [. . .]*. London: Printed for Chapman and Co., 1793.

Yahav, Amit S. *Feeling Time: Duration, the Novel, and Eighteenth-Century Sensibility*. Philadelphia: University of Pennsylvania Press, 2018.

Ylivuori, Soile. *Women and Politeness in Eighteenth-Century England: Bodies, Identities, and Power.* Routledge Studies in Eighteenth-Century Cultures and Societies. New York: Routledge, 2018.

Yolton, John W. *Thinking Matter: Materialism in Eighteenth-Century Britain.* Minneapolis: University of Minnesota Press, 1984.

Young, Sidney. *The Annals of the Barber-Surgeons of London.* London: Blades, 1890.

Zigarovich, Jolene. "Courting Death: Necrophilia in Samuel Richardson's Clarissa." *Studies in the Novel* 32, no. 2 (2000): 112–28.

———. "Egyptomania, English Pyramids and the Quest for Immortality." In *Ancient Egypt in the Modern Imagination: Art, Literature and Culture.* London: Bloomsbury, 2020.

———. "Preserved Remains: Embalming Practices in Eighteenth-Century England." *Eighteenth-Century Life* 33, no. 3 (2009): 65–104.

Zigarovich, Jolene, ed. *Sex and Death in Eighteenth-Century Literature.* Routledge Studies in Eighteenth-Century Literature. New York: Taylor and Francis, 2013.

Zohn, Kristen Miller. "Tokens of Imperfect Affection: Portrait Miniatures and Hairwork in Sense and Sensibility." *Persuasions On-Line* 32, no. 1 (Winter 2011).

INDEX

Page numbers in italics refer to figures.

ACKNOWLEDGMENTS

My deep thanks and appreciation to the editors and staff at the University of Pennsylvania Press, to include editor-in-chief Walter Biggins, Ashley Kole, and former assistant Zoe Kovacs. Brian Ostrander of Westchester Publishing Services and the production team for Penn Press handled the manuscript with great care and attention to detail. I am especially grateful to Jerome Singerman, who initially supported my project and guided its direction for several years. Without his editorial acumen, this book would not have its current scope, framework, and vision. I also owe a huge debt to Paul Saint-Amour for his encouragement throughout this process. I must thank Daniel O'Quinn for suggesting Penn Press and connecting me with Jerry.

From its inception, the project was informed and closely critiqued by Robert Folkenflik. His constant input and suggestions over the years were invaluable. Bob's mentorship, friendship, and unflagging support are dearly missed. I am a better scholar for having known him.

I must thank Marc Redfield, my advisor, whose courses and conversations inspired my interests in death studies and the novel. His brilliance and scholarly rigor taught me an incredible amount about the challenges of literary research. I owe much to other current and former Claremont Graduate University professors, colleagues, and students, including Lori Anne Ferrell, Wendy Martin, Maureen O'Connor, Melissa Bishop, and Emily Schuck, my stalwart editorial assistant. Jeremy Chow's humor and support bolstered me throughout the research and writing process.

This book would not have its current shape without the detailed and insightful suggestions of the anonymous readers, as well as input from generous colleagues. My special thanks to Marilyn Francus for her insightful responses to early drafts of the manuscript, as well as to Michelle Bloom, Emily Dowd-Arrow, and numerous ASECS members who also responded to drafts. My work on the eighteenth-century body was inspired by the brilliant work of Helen Deutsch. I'm grateful for her support and friendship.

I completed this book during a period of professional development leave from the University of Northern Iowa in the spring term of 2019. I am grateful for the support provided by the Graduate College, John Fritch, Dean of the College of Humanities and Sciences, as well as the support of my colleagues in the Department of Languages & Literatures. I must thank Jesse Swan, who has been a constant source of inspiration, support, and friendship. My thanks to mentor Jerry Klinkowitz, as well as Jeremy Schraffenberger, who supported this project from its earliest stages. I'm grateful to my UNI writing partners Elizabeth Sutton, Leisl Carr Childers, and Cara Burnidge. My former UNI graduate student Rachel Sears did an excellent job tackling another index project. Thankfully, Wendy Hoofnagle, Heather Jerónimo, Catherine MacGillivray, Jennifer McNabb, and Elizabeth Zwanziger endured desperately needed zoom wine talks during the pandemic.

Finalizing the manuscript benefited from time as a fellow in 2021 at the Institute for Advanced Studies in the Humanities, University of Edinburgh, and in 2021–2022 at the Netherlands Institute for Advanced Study, The Royal Netherlands Academy of Arts and Sciences, Amsterdam.

This book would not have been possible without the curators, archivists, and librarians who gathered materials, patiently met with me, and made numerous suggestions. I'm especially grateful to the staff at the Victoria and Albert Museum, the British Museum, the British Library, the Hunterian—University of Glasgow, and the Wellcome Institute. The V&A's Richard Edgcumbe and Rachel Church, the British Museum's Virginia Smithson, Angela Roche, Chris Coles, and Enrico Zanoni, and the Royal College of Surgeons of Edinburgh's Aaron Fleming and Steven Kerr were extremely generous with their time.

Chapter 3 is a revised version of "Preserved Remains: Embalming in Eighteenth-Century England" in *Eighteenth-Century Life* 33, 3 (2009): pp. 65–104. I gratefully acknowledge permission from Duke University Press to draw from that material.

The research for this book has been made possible in part by a summer stipend from the National Endowment for the Humanities. Any views, findings, conclusions, or recommendations expressed in this book do not necessarily represent those of the National Endowment for the Humanities.

I'm deeply grateful for the constant support of my family and friends who have listened to me talk about death, dying, and dead bodies for years. My sister June and her husband Nick have been supportive in so many ways, and their beautiful daughters, Rebecca and Shelby, inspired my research travels. My parents, Joe and Judy, have endured my frustrations and tears, but also have enjoyed my academic pleasures and achievements. I couldn't ask for more understanding, inquisitive, and proud parents.

I owe everything to my husband Steven Bamberger: archival sidekick, frequent patron, technological guru, and patient listener. No words can describe how grateful I am for your unflagging support.

Working on a book about death and material culture in the eighteenth century during a global pandemic was challenging, at times uncanny, and surprisingly cathartic. Through a painful time of struggles and loss—including grieving the death of my funny and endearing mother-in-law Dolly—examining historical mourning objects surprisingly gave me purpose. To her, this book is dedicated.

Kolibri 360 GmbH
Ferdinand-Jühlke-Straße 7
99095 Erfurt, Deutschland
produktsicherheit@kolibri360.de